Human muscle fatigue: physiological mechanisms

Human muscle fatigue: physiological mechanisms

Ciba Foundation symposium 82

1981

Pitman Medical
London

ISBN 0 272-79618-2

Published in May 1981 by Pitman Medical Ltd, London. Distributed in North America by CIBA Pharmaceutical Company (Medical Education Administration), Summit, NJ 07901, USA.

Suggested series entry for library catalogues:
Ciba Foundation symposia.

Ciba Foundation symposium 82
x + 314 pages, 84 figures, 8 tables

British Library Cataloguing in publication data:
Human muscle fatigue. – (Ciba Foundation
  symposium; 82)
  1. Muscle
  2. Fatigue
  I. Porter, Ruth   II. Whelan, Julie
  III. Series
  612'.74   QP321

Text set in 10/12 pt Linotron 202 Times, printed and bound
in Great Britain at The Pitman Press, Bath

Contents

Symposium on Human muscle fatigue: physiological mechanisms, held at the Ciba Foundation, London, 9–11 September 1980
Editors: Ruth Porter (Organizer) and Julie Whelan

Participants

B. BIGLAND-RITCHIE John B. Pierce Foundation Laboratory, 290 Congress Avenue, New Haven, Connecticut 06519, USA

E. J. M. CAMPBELL Ambrose Cardiorespiratory Unit, Department of Medicine, McMaster University, 1200 Main Street West, Hamilton, Ontario, Canada L8S 4J9

H. P. CLAMANN Department of Physiology, Medical College of Virginia, Virginia Commonwealth University, MCV Station, Richmond, Virginia 23298, USA

M. J. DAWSON Department of Physiology, University College, Gower Street, London WC1E 6BT, UK

K. W. DONALD Nant-Y-Celyn, Cloddiau, Welshpool, SY21 9JE, UK

R. H. T. EDWARDS Department of Human Metabolism, Faculty of Clinical Sciences, School of Medicine, University College, Rayne Institute, University Street, London WC1E 6JJ, UK

L. GRIMBY Department of Neurology, Karolinska sjukhuset, S-104 01 Stockholm, Sweden

L. HERMANSEN Institute of Muscle Physiology, Work Research Institutes, Gydas vei 8, PO Box 8149, Dep., Oslo 1, Norway

D. K. HILL Department of Biophysics, Royal Postgraduate Medical School, Hammersmith Hospital, Ducane Road, London W12 0HS, UK

E. HULTMAN Department of Clinical Chemistry II, Karolinska Institutet, Huddinge sjukhus, S-141 86 Huddinge, Sweden

D. A. JONES Department of Human Metabolism, Faculty of Clinical Sciences, School of Medicine, University College, Rayne Institute, University Street, London WC1E 6JJ, UK

J. KARLSSON Laboratory for Human Performance, Department of Clinical Physiology, Karolinska Hospital, S-104 01 Stockholm, Sweden

O. C. J. LIPPOLD Department of Physiology, University College, Gower Street, London WC1E 6BT, UK

P. T. MACKLEM McGill University Clinic, Royal Victoria Hospital, 687 Pine Avenue West, Montreal, Quebec, Canada H3A 1A1

P. A. MERTON Physiological Laboratory, Downing Street, Cambridge, CB2 3EG, UK

J. A. MORGAN-HUGHES The National Hospital, Queen Square, London WC1N 3BG, UK

J. MOXHAM* St James' Hospital, Sarsfeld Road, London SW12 8HW, UK

E. A. NEWSHOLME Department of Biochemistry, University of Oxford, South Parks Road, Oxford, OX1 3QU, UK

L. G. PUGH Hatching Green House, Harpenden, Herts, ALS 2JV, UK

C. ROUSSOS Meakins-Christie Laboratories, McGill University, 3775 University Street, Montreal, Quebec, Canada H3A 2B4

B. SALTIN Laboratory for the Theory of Gymnastics, August Krogh Institute, University of Copenhagen, 13 Universitetsparken, 2100 Copenhagen, Denmark

H. SJÖHOLM Department of Clinical Chemistry II, Karolinska Institutet, Huddinge sjukhus, S-141 86 Huddinge, Sweden

---

Present address: The Brompton Hospital, Fulham Road, London SW3 6HP.

J. A. STEPHENS   Sherrington School of Physiology, St Thomas's Hospital Medical School, London SE1 7EH, UK

C. M. WILES   The National Hospital, Queen Square, London WC1N 3BG, UK

D. R. WILKIE   Department of Physiology, University College, Gower Street, London WC1E 6BT, UK

Human muscle function and fatigue

R. H. T. EDWARDS

Department of Human Metabolism, University College London School of Medicine, University Street, London WC1E 6JJ, UK

Abstract Fatigue is defined as a failure to maintain the required or expected force. The force of a voluntary contraction is graded according to both the tension generated in each muscle fibre and the number of fibres recruited. The same is true of fatigue. Percutaneous electrical stimulation of a muscle via its motor nerve allows the contractile function to be measured independently of volition. Studies have been made of the forces generated isometrically at different stimulation frequencies (frequency: force curve), and of fatiguability (tendency to lose force in a given time at specified stimulation frequencies), in the quadriceps and adductor pollicis muscles. Electrical stimulation recordings of the programmed stimulation myograms distinguish forms of muscle fatigue. Low frequency fatigue which implies impaired excitation–contraction coupling is long-lasting, whereas high frequency fatigue which represents impaired muscle membrane excitation recovers rapidly.

Electromyographic (EMG) indicators of fatigue are well recognized but their use is limited because they cannot alone indicate whether alterations in excitation–contraction coupling underlie fatigue. Alterations in the power spectrum of the EMG precede (force) fatigue in sustained maximum voluntary contractions.

Fatigue may ultimately be due to a failure of the rate of energy supply to meet demand, but the precise expression of this defect may vary, such that failure of excitation or of activation may predominate over failure of the energy supply.

There are many meanings of the word 'fatigue' in use today (Table 1), but for the purpose of this symposium muscular fatigue is defined as a failure to maintain the required or expected force.

My research on the physiology of fatigue in human muscle started ten years ago when I was privileged to spend a year as a Wellcome Swedish Research Fellow at the Karolinska Institute in Stockholm. By 1970 a considerable amount had been learned, with the aid of needle biopsy (reviewed by

1981 Human muscle fatigue: physiological mechanisms. Pitman Medical, London (Ciba Foundation symposiun 82) pp 1–18

TABLE 1 Possible meanings of 'fatigue'

Definition

1. Impaired intellectual performance
2. Impaired motor performance
3. Increased EMG activity for given performance
4. Shift of EMG power spectrum to low frequencies
5. Impaired force generation

Confusion of perception associated with fatiguing muscular activity

1. Increased effort of maintaining force
2. Discomfort or pain associated with muscular activity
3. Perceived impairment of force generation

Edwards et al 1980), about metabolic changes during exercise in man. In this short review of the development of ideas of fatigue in human muscle I shall begin by considering energy metabolism before going on to more recent studies of excitatory and other electrophysiological features of fatigue. I have previously reviewed these in detail (Edwards 1979).

Energy metabolism

The studies of energy metabolism during isometric contraction of the quadriceps muscle, sustained to the point of fatigue, at different forces (Ahlborg et al 1972) offered a good opportunity to explore the metabolic factors influencing fatigue. The energy requirement as indicated by ATP turnover (rate) is increased with contraction force (Fig. 1). Endurance (that is, the time to fatigue) follows a curvilinear function (Rohmert 1960). The product of endurance time and energy turnover (rate) gives a measure of the total ATP turnover or energy cost of the contraction. It is evident that fatigue with high force contractions occurs with less energy expenditure than fatigue with long-lasting contractions at lower forces. Fatigue thus cannot be due simply to depletion of energy stores to any particular critical level. Since such isometric contractions are made under conditions that are essentially anaerobic (for details see Edwards 1976), metabolic products accumulate that might be contributing to fatigue. The greatest accumulation of lactate (+ pyruvate) occurs when forces around 50% maximum voluntary contraction are sustained to fatigue (Ahlborg et al 1972). The accumulation of lactate cannot therefore be the principal factor responsible for fatigue over the whole range of contraction forces (Karlsson et al 1975). We obtained clear evidence that fatigue could not be attributed to either simple energy depletion or product accumulation in studies in which successive isometric contractions were made

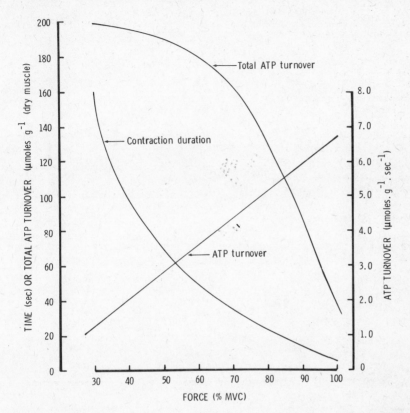

FIG. 1. Energy cost of sustaining to fatigue isometric contractions of the quadriceps muscle in normal subjects. The curve showing contraction duration is that of Rohmert (1960); the linear relation between ATP turnover (rate) and force is based on results published by Ahlborg et al (1972) and Edwards (1976). The curve for total ATP turnover is the product of the two other functions as described by Edwards (1976). (Reproduced from Rennie & Edwards 1981 with permission. © Academic Press Inc. (London) Ltd.)

to fatigue with periods of recovery but without restoration of the circulation (Fig. 2). The fact that further contractions were possible (albeit of shorter duration) while metabolic changes were progressing proved that the point of fatigue at the end of the first contraction was not solely due to metabolic factors (Edwards et al 1971). I was privileged to present this work at the excellent symposium on 'Muscle Metabolism during Exercise' held in Stockholm exactly ten years ago. My interest in fatigue continued through collaboration with Swedish colleagues in a study of the effects of altering muscle temperature on energy metabolism and fatigue (Edwards et al 1972). This study confirmed previous observations that endurance is reduced with passive cooling (due to impaired neuromuscular transmission, see below) and

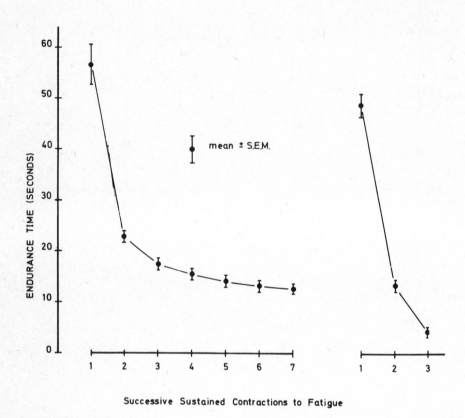

FIG. 2. Endurance for successive isometric contractions sustained to fatigue at a force of two-thirds maximum voluntary contraction (MVC) with 20 s recovery intervals with a free circulation (left) and with continuing ischaemia (right). (Combined results from two subjects, $n = 12$.) (From Edwards et al 1971.)

reduced also with warming. Endurance was greatest at a muscle temperature of about 30 °C, somewhat higher than the optimum of 27 °C reported by Clarke et al (1958). (There is, however, nothing particularly striking about this optimum temperature. It represents the 'cross-over' between two competing influences, impaired excitation and a factor related to the increase in metabolic rate with temperature.) Evidence from analysis of needle biopsy samples obtained at the start and end of repeated contractions, each sustained to fatigue at two-thirds maximum voluntary contraction (MVC), supported the suggestion (Bergström et al 1971) that the *rate* at which energy is supplied from anaerobic glycolysis might be limited by the inhibition of phosphofructokinase by the accumulated products of anaerobic glycolysis (hydrogen ions and lactate). The inhibition, as indicated by the ratio of fructose 1,6-diphosphate to fructose 6-phosphate in muscle, was greater at the highest muscle temperature when fatigue was most obvious. Apart from inhibiting glycolysis the accumulation of hydrogen ions is now thought to impede force generation by inhibiting the calcium activation of actomyosin ATPase Donaldson et al 1978, Hermansen 1979). It is worth noting here that endurance in dynamic exercise is prolonged when the subject is alkalaemic and reduced when acidaemic (N. L. Jones et al 1979), possibly for the same reason.

The question arises as to whether energy can be exchanged at a sufficiently rapid rate to permit muscular activity to continue without fatigue. This is difficult to investigate since what is being asked is whether the instantaneous rate of energy supply is adequate to meet the demand at the moment of fatigue.

Studies in patients with altered energy supply processes in muscle offer the opportunity to see whether other mechanisms are operating. Endurance is prolonged in patients with hypothyroidism and reduced in hyperthyroidism (Wiles et al 1979). With electrical stimulation at a low frequency (20 Hz) such that neuromuscular transmission was unimpaired, despite continuing activity with ischaemia there was prolonged endurance without any fading of the action potential or force loss in hypothyroid patients. Fatigue set in early and was associated with fading of the synchronous evoked potential and twitch in patients with impaired glycolysis (Wiles et al 1981). There are thus clear indications of altered membrane excitation in patients with altered muscle energy metabolism.

'Central' versus 'peripheral' fatigue

The command chain for voluntary muscular activity involves many steps and force failure—that is, fatigue—can occur as a result of impairment at any one or more link in the chain of command. As a simple practical analysis it is worth separating central fatigue from peripheral fatigue (Table 2). In the history of the study of human muscle fatigue it was popular in the early years to attribute fatigue to failure of central neural processes (Waller 1891, Mosso 1915). Comparisons between forces generated by maximum stimulated contractions and forces in maximum voluntary contraction at different stages of the experiment allowed central fatigue to be assessed (Mosso 1915, Bigland-Ritchie et al 1978). In unfatigued muscle the force of the MVC could be matched by that of a maximum stimulated contraction and it was also possible to demonstrate that the maximum rate of metabolic heat production

TABLE 2 Practical classification of muscular fatigue

Physiological mechanism	Clinical condition

Central fatigue
  Failure (voluntary or involuntary) of
  neural drive, resulting in:
    (a) reduction in number of functioning
        motor units
    (b) reduction in motor unit firing
        frequency
	Neurasthenia, hysterical paralysis and other
	  conditions in which motivation for voluntary
	  motor activity may be impaired

Peripheral fatigue
  Failure of force generation of the whole
  muscle:
  high frequency fatigue
    (a) impaired neuromuscular
        transmission
 	
    (b) failure of muscle action
        potentials
  low frequency fatigue
  impaired excitation–contraction
    coupling
	
	(a) myasthenia gravis
	    cooling of muscle
	    partial curarization
	(b) myotonia congenita
	    glycolytic disorders
	
	mitochondrial disorders
	dantrolene sodium treatment (for spasticity)
	? myotonia congenita
	? hypokalaemic periodic paralysis
	? Duchenne muscular dystrophy

was the same during maximum voluntary contractions as with maximum stimulated contractions (Edwards 1975).

During sustained MVCs of the quadriceps muscle in well-motivated subjects there is a greater tendency for central fatigue to develop as time passes, but this may be temporarily overcome in a brief 'super-effort' (Bigland-Ritchie et al 1978).

The importance of peripheral fatigue was first clearly demonstrated by Merton (1954). He showed that force generation was impaired with twitches of the adductor pollicis muscle provoked by supramaximal stimulation of the ulnar nerve at the wrist as a result of a sustained MVC, during which the muscle became fatigued such that maximum force generation was less than 20% of the force obtained with an MVC with unfatigued muscles. The same study with a sphygmomanometer cuff inflated to a pressure greater than systolic blood pressure resulted in earlier fatigue and there was no recovery of the twitch until the circulation to the muscle was restored. The action potential recorded from the surface of the muscle appeared to be unchanged in both experiments, suggesting that excitation–contraction coupling was also impaired, especially in the contraction made with ischaemia.

Fatigue due to failure of excitation or activation

The preservation of the action potential in Merton's study could be taken to
mean that neuromuscular transmission was unaffected, but this conclusion
was contested by Naess & Storm-Mathisen (1955). Impaired neuromuscular
transmission has been thought to be responsible for fatigue in sustained
MVCs of the first dorsal interosseous muscle, since in the early part of the
contraction force and surface electrical activity (smoothed rectified electromyogram) fell in proportion (Stephens & Taylor 1972). However, from
recent studies of fatiguing maximal contractions of the quadriceps it was
concluded that surface EMG provided no evidence that neuromuscular
junction failure is the limiting factor (Bigland-Ritchie et al 1978). We
concluded that, by analogy with studies of isolated curarized mouse muscles,
fatigue may be due to reduced excitability of the muscle membrane in
addition to, or perhaps more importantly than, impaired neuromuscular
transmission (Jones et al 1979). This conclusion is supported by the recent
dramatic experiments that Dr Merton will describe later (Merton et al 1981,
this volume), in which muscle is directly stimulated by brief shocks of
1000–1500 V.

A series of tests of the contractile function of human muscle have been
developed in the adductor pollicis (following the technique of Merton 1954),
in the quadriceps muscle (Edwards et al 1977b), in the sternomastoid muscle
and in the diaphragm (Moxham et al 1981). The frequency:force characteristics of human muscle have also been determined *in vitro* (Moulds et al 1977,
Faulkner et al 1979) and found to agree with those *in vivo*.

The muscle is stimulated by its motor nerve at a succession of frequencies
and a 'programmed stimulation myogram' is obtained that gives a picture of
the contractile function of the muscle and also provides a means for
identifying different types of voluntary muscle fatigue (see Table 2, p 6).
High frequency fatigue occurs when cooling impairs neuromuscular transmission and also in myasthenia gravis, and when membrane function is disturbed
in myotonia congenita (Edwards 1980). Neuromuscular transmission is also
impaired by partial curarization (Fig. 3). High frequency fatigue was
observed in studies in the cat soleus (Davis & Davis 1932) and is a feature of
normal human muscle function. It appears that in order to optimize force
maintenance during sustained voluntary contractions, the motor firing frequency is reduced (D. A. Jones et al 1979). In practice it is necessary to
reduce the stimulation frequency to simulate the time course of a maximum
voluntary contraction with an electrically stimulated contraction. The maximum firing frequency at the start of a contraction may be over 100 Hz but
during sustained activity the frequency rapidly falls to 20 Hz or less.

High frequency fatigue is thought to be a consequence of impaired

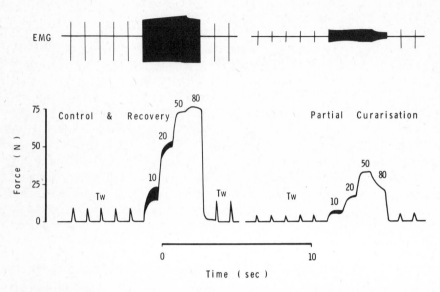

FIG. 3. Programmed stimulation of adductor pollicis showing effects of partial curarization. (Campbell et al 1977.)

membrane excitation, particularly associated with accumulation of potassium (or, conversely, depletion of sodium) (Bezanilla et al 1972) in the inter-fibre space, or more particularly in the extracellular fluid contained in the transverse tubular system (Bigland-Ritchie et al 1979).

Low frequency fatigue (Edwards et al 1977a) may be produced by a succession of contractions made under anaerobic conditions (Fig. 4). It is characterized by a selective loss of force at low stimulation frequencies despite recovery of the force generated at high stimulation frequency. This type of fatigue is not related to depletion of ATP or phosphoryl creatine and is long-lasting, taking several hours to recover. The mechanism underlying low frequency fatigue is thought to be impaired excitation–contraction coupling such that less force is generated for each individual membrane excitation. This may be partly because of reduced release of calcium or possibly because of impaired transmission in the transverse tubular system, as a result of damage in the period of ischaemic activity. Studies in rat fast and slow muscle (Kugelberg & Lindegren 1979) support the above explanation and further suggest that resistance to the development of low frequency fatigue is a particular characteristic of fibres with a high oxidative capacity.

Low frequency fatigue has been found in the sternomastoid of normal subjects and patients after loaded breathing and sustained maximum voluntary ventilation and in the diaphragm after similar manoeuvres (Moxham et al 1981). It is evident (Table 2) in some metabolic myopathies (Wiles et al 1981).

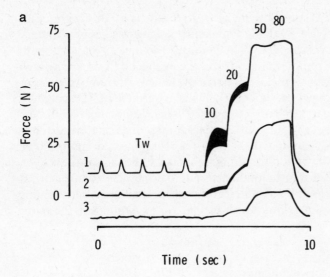

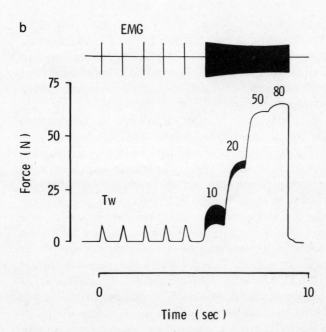

FIG. 4. Programmed stimulation of adductor pollicis. (a) Immediate effects of ischaemic fatigue. 1. Control. 2. Midway through series of contractions. 3. Immediately after ischaemic contractions. (b) Recovery 1 h 50 min after ischaemic fatigue, illustrating low frequency fatigue. A 20 Hz tetanus is reduced to 81% of the control value (a), while an 80 Hz tetanus is 98% of control. No change was seen in the evoked action potential.

There are quite different consequences of these types of fatigue with voluntary muscular activity. With high frequency fatigue maximum voluntary contractions cannot be sustained. The tendency to fall in patients with myasthenia gravis or myotonia congenita may be explained by persisting high frequency fatigue such that muscles are not activated with the high motor unit firing frequencies involved in sudden (ballistic) contractions (Desmedt & Godaux 1977), necessary to correct balance when stumbling. With low frequency fatigue there is a need in submaximal contractions (involving motor firing frequencies in the range of 10–30 Hz) for recruitment of more motor units or an increase in the motor unit firing frequency if the same force is to be sustained as previously or if further fatigue is not to occur. The total electrical activity of the smoothed rectified electromyogram is increased in this situation but this does not tell us which of the two adaptations is the more important.

Electromyographic changes in fatigue

The electromyogram has been used as a guide to fatigue for many years (Edwards & Lippold 1956). In fatigue the smooth rectified electromyogram (or 'integrated' EMG) tends to be increased in relation to the force of contraction (Kuroda et al 1970). This may be due to one or more of several factors including impaired excitation–contraction coupling, increased firing frequency, and/or synchronization of motor unit recruitment. An approach based on the smooth rectified EMG is however of limited value as an objective guide to fatigue. An alternative that has been gaining in popularity is based on analysis of the power spectrum of the crude electromyogram recorded from the surface of the muscle. This approach has been applied to the study of fatigue in the quadriceps (Komi & Tesch 1979) and in the respiratory muscles (Grassino et al 1979). The sign of fatigue is a shift of the power spectrum towards low frequencies and the simple index used is a reduction in the ratio of power in high to low frequencies (i.e. a fall in the high:low ratio). The question is whether this is a measure of fatigue as defined in terms of force generation. It was a particular pleasure for Dr Moxham and I to work with Dr Macklem, Dr Roussos and their colleagues in Montreal in autumn 1978 in order to measure the changes in the power spectrum occurring in fatiguing contractions in which the contractile properties had been defined with electrical stimulation techniques for recording the programmed stimulation myogram (described above).

When we measured the power spectrum during sustained maximum voluntary contraction of the quadriceps muscle we saw an early, rapid fall in the ratio (at a time when there was little or no change in force). Later in the

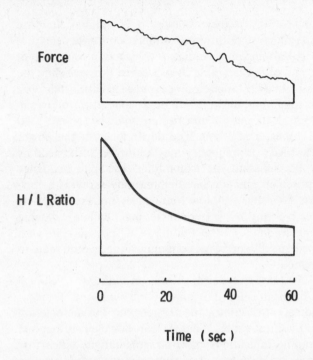

Force

H / L Ratio

0   20   40   60

Time (sec)

FIG. 5. Change in high/low ratio (130–233 Hz/20–40 Hz) of EMG power spectrum during 60 s maximum voluntary contraction of quadriceps muscle.

contraction, when force was declining, there was no change in the high:low ratio (Fig. 5). We then found that on repeating earlier experiments (Edwards et al 1971) in which contractions were made with ischaemic recovery between contractions, there was recovery of force and of the high:low ratio at a time when lactate concentrations were increasing in the muscle. This indicated that the alteration in high:low ratio could not be simply attributed to accumulation of lactate, as Mortimer et al (1970) had previously suggested. This conclusion is further supported by our observations, to be discussed later by Dr Wiles, in which alterations in high:low ratio are also seen in patients with glycolytic disorders (Wiles et al 1981, this volume). It thus seems that the alteration in high:low ratio is a consequence of muscular activity and occurs early, but it does not appear to correspond directly with the development of fatigue. The alteration in the high:low ratio of the power spectrum cannot be simply attributed to the development of low frequency fatigue (Moxham et al 1979), either. After a succession of ischaemic contractions of the quadriceps muscle such that the ratio of force at 20 Hz to force at 100 Hz was reduced to approximately 50% of the value for unfatigued muscle, there was no alteration in the high:low ratio. In these studies it is notable that the measurement of contractile function at different stimulation frequencies and of the power spectrum with sub-maximal voluntary contraction was made at least 10 minutes after the end of the fatiguing contractions, thus avoiding the immediate (acute) effects of muscle fatigue on neuromuscular transmission.

The shift in power spectrum is still unexplained but it appears to follow the time course of the reduction in the motor unit firing frequency (as expected from the study of D. A. Jones et al 1979). The dependence of the power spectral shift of the muscle EMG on frequency may be further influenced by impaired neuromuscular transmission, as exemplified by high frequency fatigue when the neuromuscular junction/sarcolemma may serve as a low-frequency band pass filter. Thereby, only low frequencies can excite muscle contraction, giving an effective shift to low frequencies independently of what may be the central driving frequency. The finding of a spectral shift with partial curarization (Dr D. Pengelly, personal communication) would seem to support this view if confirmed.

The usefulness of the EMG power spectral shift as an indicator of clinical muscle fatigue has yet to be established. Fortunately the shift tends to occur earlier than any clear indication of contractile fatigue, but the dissociation may be too broad to be of practical value. What is clear, however, is that one cannot use electromyographic indications of contractile fatigue since that approach does not take account of impaired excitation–contraction coupling as a possible mechanism of fatigue, and in this respect is not a 'fail-safe' indicator of clinical muscle fatigue. There is a logical necessity, therefore, to adhere to a definition of fatigue based on force generation.

Fibre types and fatigue

The considerations of fatigue in terms of altered energy supply for contractile processes, membrane excitation or the susceptibilities of muscle cells to high or low frequency fatigue are based on *in vivo* studies of the muscle as a whole. With increasing voluntary contractions there is progressive recruitment of slow twitch, fatigue-resistant fibres at low forces with addition of high threshold (high frequency) fast twitch, fatiguable fibres for greater contractions (Garnett et al 1978). Force at any time during contraction will depend on the number of active fibres and on the force generated by each fibre. Human muscles comprise fibres of different fibre types with quite distinct metabolic and contractile characteristics. In particular, there are differences in the fatiguability of the two fibre types, the fast twitch glycolytic (Type II) muscle fibres being more fatiguable than the slow twitch oxidative (Type I) muscle fibres. Studies of human muscle *in vitro* (Faulkner et al 1979) indicate that muscles with a high proportion of Type II fibres tend to fatigue more rapidly than those with a high proportion of the oxidative Type I fibres. Thus the rectus abdominis, which has a high proportion of Type II fibres, fatigues much more rapidly than the diaphragm or intercostal muscles. The shift in EMG power spectrum was found to be greater (as was fatigue) in subjects with a high proportion of fast twitch glycolytic fibres (Komi & Tesch 1979). Whether this is primarily because of a reduction in the recruitment of high threshold motor units (central fatigue) or a failure of excitation of the sarcolemma is not clear. The likelihood of a close correlation between the function (fatiguability) of motor neuron and muscle cell, as suggested previously (D. A. Jones et al 1979, Kugelberg & Lindegren 1979), is however put in doubt by the finding in recent cross-innervation experiments (in cats) that fast and slow twitch muscles retained their characteristic fatigue resistance properties regardless of whether their innervation had previously supplied a fatiguable or fatigue-resistant muscle (Edgerton et al 1980).

Interrelation of metabolic and electrical factors in fatigue

For individual muscle cells as well as for the muscle as a whole there is a very close relationship between excitatory/activation processes and energy metabolism. Failure of one will affect the other and will in turn result in force fatigue. It is suggested, for the sake of argument, that these interrelationships may be visualized with a three-dimensional diagram (Fig. 6). Force is related to excitation/activation by the frequency:force relationship found for human muscle (Edwards 1977b). Force was found in frog muscle, studied by phosphorus nuclear magnetic resonance spectroscopy, to be directly proportional to the rate of ATP hydrolysis (Dawson et al 1978). For the present purposes it is assumed that an analogous relationship exists for human muscle. The relationship between excitation/activation and energy metabolism has been studied in single frog muscle fibres (Nassar-Gentina et al 1978) but though the findings agree with our observations on low frequency fatigue in man (Edwards et al 1977a), no mathematical relation is available which would be applicable in the development of Fig. 6. Clearly, there will be impaired excitation/activation if the energy supply is critically low. Conversely, if there are other causes of impaired excitation/activation, such as accumulation of potassium in the extracellular space followed by impaired membrane excitation and/or transverse tubular function, possibly associated with Na^+ depletion (Bezanilla et al 1972) and leading to impaired excitation–contraction coupling, there will be a reduced demand for energy for the contractile mechanism. There thus appears to be a 'fail-safe' mechanism which prevents the muscle going into rigor through depletion of ATP, as

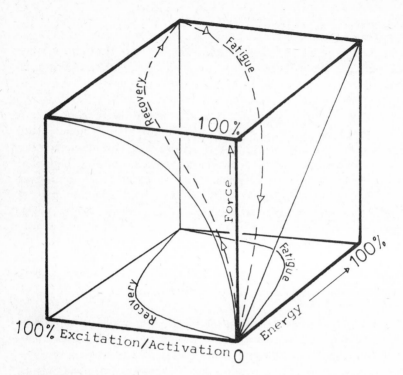

FIG. 6. Three-dimensional plot of possible relation between force and energy (ATP hydrolysis, Dawson et al 1978) and excitation/activation (frequency:force curve, Edwards et al 1977b). This representation is intended to emphasize our present lack of quantitative information about the precise interrelationships between the variables. Recovery may follow a different path, as a result of the different time courses of recovery processes. Both fatigue and recovery pathways may vary according to the type of muscular activity.

suggested by Nassar-Gentina et al (1978) and Kugelberg & Lindegren (1979). Perhaps the same mechanism is operating, though less effectively, in patients with myophosphorylase deficiency (Wiles et al 1981), since excitation/activation may fail and yet a contracture develop. Analysis of such muscle in contracture did not reveal low ATP concentrations such as might be expected in rigor (Rowland et al 1965). The relation between energy metabolism and excitation/activation must therefore remain hypothetical, but at any one time during the critical period when force is failing, there may be a predominant influence by one or other of the principal mechanisms (Table 3), resulting in a particular or characteristic pathway in the three-dimensional plot. It is also likely that the return or recovery pathway will be different from the fatigue pathway, since excitation not only recovers very quickly but can also recover under anaerobic conditions (except in patients with myophosphorylase de-

TABLE 3 Metabolic and electrophysiological consequences of muscular activity which might lead to fatigue

1. Depletion of energy for contractile mechanism limiting rate of ATP supply
2. Impaired energy supply for membrane function, e.g.: impaired generation and propagation of sarcolemmal action potential; impaired calcium pumping by sarcoplasmic reticulum
3. Accumulation of products

Intracellular H+	(a)	may inhibit activity of phosphofructokinase and phosphorylase
	(b)	may reduce Ca^2-activated actomyosin interaction
Extracellular K+	(a)	impaired sarcoplasmic action potential generation and propagation
(depletion of	(b)	impaired action potential in transverse tubular system resulting in
Na+)		reduced efficiency of excitation–contraction coupling

ficiency: Edwards et al 1981) when there is little or no recovery of phosphoryl creatine in muscle (Harris et al 1976).

When we consider the ways in which human muscle may fail to generate the required force, with a resulting loss in performance and altered perception of muscular proprioception (Table 1), it is evident that there are many factors that are playing a role and may fail. What is increasingly evident as a result of our own research of the past ten years is that though energy metabolism is undoubtedly of considerable interest there are possibly more important alterations in excitation/activation which may override it. Such considerations are particularly important in studies in patients with altered energy metabolism or impaired membrane function (as in myotonia congenita). The importance of studying patients with selected disorders of energy metabolism or membrane function is that they offer the opportunity to analyse particular sections of the chain of command for muscular contraction in a way that is analogous to the historically important use *in vitro* of enzyme inhibitors. The approach described, and the study of human muscle fatigue in general, is also potentially important for the development of drugs affecting human skeletal muscle. The gross alterations in skeletal muscle function seen in conditions of experimental muscular fatigue may be considerably more extensive than the pathological alterations, with the advantage of reversibility, whether rapid or more gradual.

I hope that this symposium will result in our gaining an integrated view of the mechanisms underlying skeletal muscle fatigue in man, not only for their potential interest and importance for athletic performance, but also for their particular relevance to understanding and treating human myopathy.

Acknowledgements

Support from The Wellcome Trust and Muscular Dystrophy Group of Great Britain is gratefully acknowledged.

REFERENCES

Ahlborg B, Bergström J, Ekelund LG, Guarnieri G, Harris RC, Hultman E, Nordesjö L-O 1972 Muscle metabolism during isometric exercise performed at constant force. J Appl Physiol 33:224-228

Bergström J, Harris RC, Hultman E, Nordesjö L-O 1971 Energy rich phosphagens in dynamic and static work. Adv Exp Med Biol 11:341-355

Bezanilla F, Caputo C, Gonzalez-Serratos H, Venosa RA 1972 Sodium dependence of the inward spread of activation in isolated twitch muscle fibres of the frog. J Physiol (Lond) 223:507-523

Bigland-Ritchie B, Jones DA, Hosking GP, Edwards RHT 1978 Central and peripheral fatigue in sustained maximum voluntary contractions of human quadriceps muscle. Clin Sci Mol Med 54:609-614

Bigland-Ritchie B, Jones DA, Woods JJ 1979 Excitation frequency and muscle fatigue: electrical responses during human voluntary and stimulated contractions. Exp Neurol 64:414-427

Campbell EJM, Edwards RHT, Hill DK, Jones DA, Sykes MK 1977 Perception of effort during partial curarization. J Physiol (Lond) 263:186-187P

Clarke RSJ, Hellon RF, Lind AR 1958 The duration of sustained contractions of the human forearm at different temperatures. J Physiol (Lond) 143:454-473

Davis H, Davis PA 1932 Fatigue in skeletal muscle in relation to the frequency of stimulation. Am J Physiol 101:339-356

Dawson MJ, Gadian DG, Wilkie DR 1978 Muscular fatigue investigated by phosphorus nuclear magnetic resonance. Nature (Lond) 274:861-866

Desmedt JE, Godaux E 1977 Ballistic contractions in man: characteristic recruitment pattern of single motor units of the tibialis anterior muscle. J Physiol (Lond) 264:673-693

Donaldson SK, Hermansen L, Bolles L 1978 Differential, direct effects of H^+ on Ca^{2+}-activated force of skinned fibers from the soleus, cardiac and adductor magnus muscles of rabbits. Pflügers Arch Eur J Physiol 376:55-65

Edgerton VR, Goslow GE Jr, Rasmussen SA, Spector SA 1980 Is resistance of a muscle to fatigue controlled by its motor neurones? Nature (Lond) 285:589-590

Edwards RHT 1975 Muscle fatigue. Postgrad Med J 51:137-143

Edwards RHT 1976 Metabolic changes during isometric contraction of the quadriceps muscle. Thermodynamics of muscular contraction in man. Med Sport (Basel) 9:114-131

Edwards RHT 1979 Physiological and metabolic studies of the contractile machinery of human muscle in health and disease. Phys Med Biol 24:237-249

Edwards RHT 1980 Studies of muscular performance in normal and dystrophic subjects. Br Med Bull 36:159-164

Edwards RG, Lippold OCJ 1956 The relation between force and integrated electrical activity in fatigued muscle. J Physiol (Lond) 132:677-681

Edwards RHT, Nordesjö L-O, Koh D, Harris RC, Hultman E 1971 Isometric exercise—factors influencing endurance and fatigue. Adv Exp Med Biol 11:357-360

Edwards RHT, Harris RC, Hultman E, Kaijser L, Koh D, Nordesjö L-O 1972 Effect of temperature on muscle energy metabolism and endurance during successive isometric contractions, sustained to fatigue, of the quadriceps muscle in man. J Physiol (Lond) 220:335-352

Edwards RHT, Hill DK, Jones DA, Merton PA 1977a Fatigue of long duration in human skeletal muscle after exercise. J Physiol (Lond) 272:769-778

Edwards RHT, Young A, Hosking GP, Jones DA 1977b Human skeletal muscle function: description of tests and normal values. Clin Sci Mol Med 52:283-290

Edwards RHT, Young A, Wiles CM 1980 Needle biopsy of skeletal muscle in the diagnosis of myopathy and the clinical study of muscle function and repair. N Engl J Med 302:261-271

Edwards RHT, Wiles CM, Gohil K, Krywawych S, Jones DA 1981 Energy metabolism in human myopathy. In: Schotland DL (ed) Disorders of the motor unit. Houghton Mifflin, Boston, in press

Faulkner JA, Jones DA, Round JM, Edwards RHT 1979 Contractile properties of isolated human muscle preparations. Clin Sci (Oxf) 57:20P

Garnett RAF, O'Donovan MJ, Stephens JA, Taylor A 1978 Motor unit organization of the human medial gastrocnemius. J Physiol (Lond) 287:33-43

Grassino A, Gross D, Macklem PT, Roussos C, Zagelbaum G 1979 Inspiratory muscle fatigue as a factor limiting exercise. Bull Eur Physiopathol Respir 15:105-111

Harris RC, Edwards RHT, Hultman E, Nordesjö L-O, Nylind B, Sahlin K 1976 The time course of phosphorylcreatine resynthesis during recovery of the quadriceps muscle in man. Pflügers Arch Eur J Physiol 367:137-142

Hermansen L 1979 Effect of acidosis on skeletal muscle performance during maximal exercise in man. Bull Eur Physiopathol Respir 15:229-238

Jones DA, Bigland-Ritchie B, Edwards RHT 1979 Excitation frequency and muscle fatigue: mechanical responses during voluntary and stimulated contractions. Exp Neurol 64:401-413

Jones NL, Sutton JR, Taylor R, Toews CJ 1979 Effect of pH on cardiorespiratory and metabolic responses to exercise. J Appl Physiol Respir Environ Exercise Physiol 43:959-964

Karlsson J, Funderburk CF, Essén B, Lind AR 1975 Constituents of human muscle in isometric fatigue. J Appl Physiol 38:208-211

Komi PV, Tesch P 1979 EMG frequency spectrum, muscle structure and fatigue during dynamic contractions in man. Eur J Appl Physiol 42:41-50

Kugelberg E, Lindegren B 1979 Transmission and contraction fatigue of rat motor units in relation to succinate dehydrogenase activity of motor unit fibres. J Physiol (Lond) 288:285-300

Kuroda E, Klissouras V, Milsum JH 1970 Electrical and metabolic activities and fatigue in human isometric contraction. J Appl Physiol 29:358-367

Merton PA 1954 Voluntary strength and fatigue. J Physiol (Lond) 123:553-564

Merton PA, Hill DK, Morton HB 1981 Indirect and direct stimulation of fatigued human muscle. This volume p 120–126

Mortimer JT, Magnusson R, Petersén I 1970 Conduction velocity in ischemic muscle: effect on EMG frequency spectrum. Am J Physiol 219:1324-1329

Mosso A 1915 Fatigue, 3rd edn. Drummond M, Drummond WG (transl) Allen & Unwin, London, p 334

Moulds RFW, Young A, Jones DA, Edwards RHT 1977 A study of the contractility, biochemistry and morphology of an isolated preparation of human skeletal muscle. Clin Sci Mol Med 50:291-297

Moxham J, De Troyer A, Farkas G, Macklem PT, Edwards R, Roussos C 1979 Relationship of EMG power spectrum (PS) with low and high frequency fatigue in human muscle. Physiologist 22:91

Moxham J, Wiles CM, Newham D, Edwards RHT 1981 Respiratory muscle fatigue. This volume p 197–205

Naess K, Storm-Mathisen A 1955 Fatigue of sustained tetanic contractions. Acta Physiol Scand 34:351-366

Nassar-Gentina V, Passonneau JV, Vergara JL, Rapaport SI 1978 Metabolic correlates of fatigue and of recovery from fatigue in single frog muscle fibers. J Gen Physiol 72:593-606

Rennie MJ, Edwards RHT 1980 Carbohydrate metabolism in skeletal muscle. In: Dickens F et al (eds) Carbohydrate metabolism and its disorders, vol 111. Academic Press, New York & London

Rohmert W 1960 Ermittlung von Erholungspausen für statische Arbeit des Menschen. Int Z Angew Physiol Einschl Arbeitsphysiol 18:123-164

Rowland LP, Araki S, Carmel P 1965 Contracture in McArdle's disease. Arch Neurol 13:541-544

Stephens JA, Taylor A 1972 Fatigue of maintained voluntary muscle contraction in man. J Physiol (Lond) 220:1-18

Waller AD 1891 The sense of effort: an abjective study. Brain 14:179-249

Wiles CM, Jones DA Edwards RHT 1981 Fatigue in human metabolic myopathy. This volume p 264-277

Wiles CM, Young A, Jones DA, Edwards RHT 1979 Muscle relaxation rate, fibre-type composition and energy turnover in hyper- and hypothyroid patients. Clin Sci (Oxf) 57:375-384

Glycolytic and oxidative energy metabolism and contraction characteristics of intact human muscle

ERIC HULTMAN, HANS SJÖHOLM, KENT SAHLIN and LARS EDSTRÖM*

*Department of Clinical Chemistry II, Huddinge sjukhus, S-141 86 Huddinge, Sweden and
Department of Neurology, Karolinska sjukhuset, S-104 01 Stockholm, Sweden

Abstract It is proposed that glycolytic rate may be measured as lactate accumulation after electrical stimulation of the quadriceps femoris muscle under anaerobic conditions. The ratio of glucose 6-phosphate to lactate is an internal monitor of the glycolytic pathway. The phosphocreatine/lactate ratio links glycolysis and the creatine kinase reaction and could be used to distinguish abnormalities in energy metabolism. The rate of resynthesis of phosphocreatine after stimulation when the circulation is restored should be a measure of oxidative phosphorylation. The relaxation rate seems to be a mechanical index of the metabolic state of the muscle.

Studies of energy metabolism in human muscle are a necessary prerequisite for understanding muscle function both in normal man and in patients with different disorders of the locomotor system. The estimation of energy metabolism in intact human muscle was limited to indirect methods until the mid 1960s. The utilization of different substrates during work was estimated by measurements of oxygen consumption and CO_2 production and calculation of the whole-body respiratory quotient. The uptake of substrates and release of metabolites by working muscles could also be measured, by means of catheterization of blood vessels. The introduction of the muscle biopsy technique by Bergström and Hultman around 1965 (see Bergström 1975) opened up new possibilities of directly measuring the metabolic state of human muscle in various functional situations. In a comparatively short time information accumulated on the turnover of high-energy phosphates and the use of different substrates for energy metabolism in relation to dynamic and

1981 Human muscle fatigue: physiological mechanisms. Pitman Medical, London (Ciba Foundation Symposium 82) p 19-40

isometric contractions. It was shown in these studies that the muscle glycogen store was depleted during prolonged dynamic exercise, and that the depletion was related in time to the point of exhaustion (Bergström & Hultman 1967, Hultman 1967). It could also be shown that if the glycogen content in the muscles was varied by dietary manipulations (Bergström & Hultman 1966), the endurance time for exhaustive work covaried with the size of the glycogen store (Bergström et al 1967, Hultman & Bergström 1973).

Further studies revealed that the glycogen store in the liver was also used during dynamic exercise and that this store too could be changed by dietary regimens (Hultman & Nilsson 1971, 1973). A small glycogen store in the liver resulted in a decreased blood sugar concentration during hard prolonged exercise and could limit performance capacity through lack of substrate for the central nervous system.

Thus fatigue in dynamic exercise was clearly related to lack of stored carbohydrate substrate for energy production, either in the working muscles or in the liver, or in both. This dependence on stored carbohydrate was observed at work loads corresponding to 70–85% of the subject's maximal oxygen uptake capacity (Bergström & Hultman 1972).

At low work loads the rate of utilization of glycogen is low and the whole store is not always used up at the point of exhaustion. At work loads near or above the individual's maximum oxygen uptake capacity, exhaustion is reached before the glycogen store is depleted (Saltin & Karlsson 1971). Consequently, other factors than lack of carbohydrate substrate will limit work capacity and cause fatigue at these work loads.

Measurements of the phosphagen content in muscle during dynamic exercise with different work loads showed that the phosphocreatine level during exercise was inversely related to the load (Hultman et al 1967). At high work loads producing phosphocreatine levels below 10% of the normal value at rest a significant decrease in ATP level was also observed. This decrease amounted to about 40% when the phosphocreatine store was practically depleted. At supramaximal work loads the content of phospho-creatine approached zero and this decrease seemed to determine the point of exhaustion (Hultman et al 1967).

Further studies with isometric exercise showed that during voluntary contractions with a work load close to the maximum, about 50% of the energy utilized was derived from degradation of energy-rich phosphagens and about 50% from anaerobic glycolysis utilizing the glycogen store (Bergström et al 1971). It was suggested in this study that accumulation of lactic acid during the exercise could produce a decrease in pH in the cell large enough to decrease the rate of glycolysis at the level of phosphofructokinase. A decreased rate of glycolysis could explain the decrease in ATP seen at the end of exercise and possibly be the reason for a lowering of the rate of ATP resynthesis with resultant lack of available $\sim$ P, which was suggested as the reason for inability to continue muscle contraction. These earlier studies serve as a background for a model developed by our group for testing human muscle function, to be presented below.

The aim of the model is to estimate quantitatively the different pathways of energy metabolism and relate them to the contractile behaviour of the muscle, especially features of fatigue such as decline in tension and changes in relaxation rate. The routes of ATP production that will be considered are glycolysis, oxidative phosphorylation, the creatine kinase reaction and the turnover of the adenine nucleotide pool itself.

Electrical stimulation of intact muscle

In order to standardize the activation of the muscle we have used electrical stimulation of the quadriceps femoris muscle in essentially the same way as described earlier by Edwards' group (Edwards et al 1977). Two relatively large aluminium foil electrodes (12×12 cm) were placed proximally and distally on the antero-lateral aspect of the thigh and connected to a square-wave pulse generator. To overcome the resistance of the skin and sub-cutaneous fat layer a potential of around 50 volts is needed to drive the muscle. Pulses of 50μs duration are used which most probably activate the nerve endings within the muscle tissue. The muscle force produced is recorded isometrically with the subject seated in a straight-backed chair. The operating voltage is set so that approximately 30% of the subject's maximum voluntary contraction force is produced by the knee extensors. With this strength of stimulation the superficial layers of the muscle closest to the electrodes are stimulated while the deeper parts are at rest.

This mode of muscle activation has several advantages. Voluntary contractions of the muscle could be compared to contractions elicited by direct electrical stimulation in order to distinguish between lack of central drive and impairment within the muscle in effort syndromes. Steady maximal and submaximal tetanization can be produced, which allows us to monitor a decline in tension during continued stimulation and to measure relaxation rates. The relation between stimulation frequency and tension development could also be studied. Problems with varying motor unit recruitment during voluntary contractions should be minimized during electrical stimulation, aiding the interpretation of metabolic data when biopsies are taken.

Anaerobic energy metabolism

The circulation in the thigh can be arrested by using a tourniquet. This means that no oxidation or transport of metabolites produced in the muscle cells during contraction will occur; that is, the muscle is a closed system. All products of metabolism will thus remain within the muscle and can be analysed in biopsy samples. The energy sources available in this situation are glycogen, phosphocreatine and the adenine nucleotide pool.

Methods

In a series of experiments on healthy volunteers of both sexes the anaerobic routes of energy production were measured and compared to the contractile behaviour of the muscle. A tourniquet was inflated above arterial pressure around the proximal part of the thigh to create an anaerobic situation. The muscle was stimulated with frequencies of 5–50 Hz, showing varying frequency–tension patterns in different individuals. Maximum tension was achieved at the highest frequency (50 Hz) but the tension showed a rapid decline starting after only 20 s of stimulation, probably as a result of failure of neuromuscular transmission. A lower frequency (20 Hz) was chosen for stimulation with time periods of 12, 25, 50 and 75 s respectively. Stimulation at 20 Hz gives rise to a fused tetanus producing approximately 75% of the maximum tension achievable at 50 Hz. Immediately after cessation of stimulation a biopsy was taken from the activated part of the muscle with the tourniquet still inflated. The biopsies were taken with the Radner forceps (Radner 1962) since a biopsy needle would probably have passed through the stimulated muscle layers. The biopsy material was frozen as quickly as possible in freon maintained at its melting point by liquid nitrogen and later analysed by enzymic methods for ATP, ADP, AMP, phosphocreatine, free creatine, pyruvate, lactate, glucose 1-phosphate, glucose 6-phosphate (G-6-P) and fructose 6-phosphate (Harris et al 1974) and by high pressure liquid chromatography for AMP and IMP (Sahlin et al 1978).

Glycolysis

The rate of glycolysis was expressed as the accumulation of glycosyl units from glycogen (= sum of Δ glycolytic intermediates + Δ lactate/2) or as accumulation of lactate (Fig. 1). There was a constant proportionality between the accumulation of total glycolytic intermediates and the accumulation of lactate, irrespective of the glycolytic rate. The G-6-P content rose to more than 20 μmol g^{-1} dry muscle when lactate increased to 100 μmol g^{-1} and the ratio G-6-P/lactate was always around 0.2 (see Fig. 1). The explanation for this proportionality in the glycolytic pathway when the system is closed could be that glycogen phosphorylase is rate generating and that the rest of glycolysis is

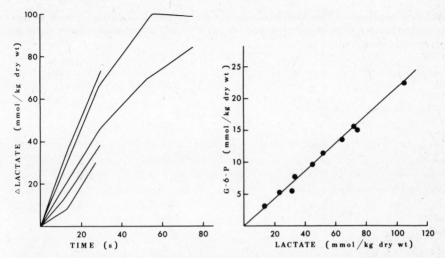

FIG. 1. Accumulation of lactate in biopsy samples from the quadriceps femoris muscle of five volunteers during electrical stimulation at 20 Hz. The blood supply to the muscle was occluded (left). The relation of glucose 6-phosphate to lactate in the same biopsy samples (right).

in a steady state. This would then mean either that phosphofructokinase does not control the rate of glycolysis in this situation, possibly due to increases in concentration of activators such as fructose 6-phosphate, AMP and NH_4^+ and decreases of deactivators such as ATP and phosphocreatine, or that phosphofructokinase exerts control over the glycolysis rate via feedback by G-6-P on glycogen phosphorylase *b*. This feedback control must be instantaneous and sensitive since no excess accumulation of G-6-P is seen when glycolysis slows down. *The constancy in the G-6-P/lactate quotient in normal subjects could be of practical use as an internal monitor of the glycolytic pathway to distinguish pathological states such as McArdle's disease or diseases with impaired phosphofructokinase activity.*

The rate of glycolysis was maximal between 12 s and 25 s of continued stimulation. Thereafter there was a progressive decline in the glycolytic rate (see Table 1). However, the integrated glycolytic rate during 50 s stimulation can be used as a measure of the glycolytic rate at this stimulation frequency. The maximum glycolytic rate was of the same order of magnitude as that observed previously during voluntary isometric contraction with a load of 95% of maximal voluntary contraction (MVC) (Bergström et al 1971).

Phosphagen utilization

The rate of degradation of phosphocreatine reached maximum values within the first 12s of stimulation and the phosphocreatine store was used com-

TABLE 1 Rate of total ATP production and relative contribution of available energy stores during electrical stimulation of the quadriceps femoris muscle as calculated from biopsy samples taken at the end of the indicated time intervals

Time interval (s)	Rate of ATP production ($\mu mol\ g^{-1}\ s^{-1}$)	% ATP derived from phosphocreatine	% ATP derived from glycolysis	% ATP derived from TAN[a]
Subject A				
0–28.6	6.08	41	57	2
28.6–55	2.12	0	91	9
55–75	0.23	0	0	100
Subject B				
0–28.5	5.02	48	48	4
28.5–52	1.62	0	97	3
52–75	1.15	0	84	16
Subject C				
0–12.6	3.63	72	26	0
12.6–25.6	3.92	41	59	0
Subject D				
0–10.7	4.79	66	33	1
10.7–29.6	3.11	31	69	2
Subject E				
0–13.8	8.54	52	47	0
13.8–29.7	4.23	15	82	4

[a]TAN, total adenine nucleotide pool (ATP + ADP + AMP).

pletely during 25–50 s of stimulation (Fig. 2). An individual variation in phosphocreatine degradation rate was found which seemed to be a reflection of the degree of tetanization at 20 Hz (Fig. 2). It has previously been shown that there is no resynthesis of phosphocreatine after contraction to fatigue unless the circulation is restored (Harris et al 1976). The explanation is presumably that glycolysis is inhibited by lactic acidosis in this situation. However, during 12–25 s of stimulation glycolysis is clearly not inhibited and hence resynthesis of phosphocreatine and continued glycolysis could occur in the time between the end of contraction and freezing of the biopsy sample, resulting in an underestimation of the rate of utilization of phosphocreatine and an over-estimation of the glycolytic rate during contraction. Harris et al (1977) showed that there is a constant exponential relationship between the amounts of phosphocreatine and lactate in biopsy samples taken immediately after termination of voluntary contractions irrespective of the type, intensity and duration of the preceding exercise. The muscle blood flow was not arrested in this series of experiments. During electrical stimulation with arrested blood

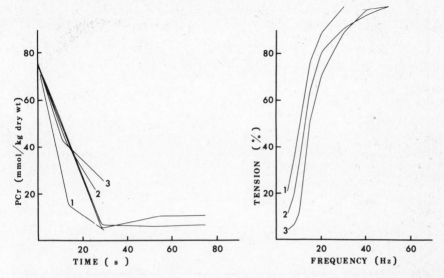

FIG. 2. The phosphocreatine (PCr) content of the quadriceps femoris muscle of five volunteers during electrical stimulation at 20 Hz. The blood circulation to the muscle was occluded (left). Frequency–tension relation during electrical stimulation for three volunteers (right). Numbers indicate the same subjects.

flow there is also an exponential relationship between phosphocreatine and lactate (Fig. 3). This relationship differs, however, from that seen during the voluntary contractions in that there is a greater decrease in phosphocreatine for a given amount of lactate accumulated. It seems instead to follow the curve obtained in resting muscle during prolonged circulatory occlusion (60–150 min of occlusion during surgical knee operations: H. Sjöholm, K. Sahlin, J. Larsson, & E. Hultman, unpublished observations).

The difference in the phosphocreatine/lactate ratio between voluntary contraction with intact muscle blood flow on the one hand and electrical stimulation and resting metabolism during occlusion on the other is probably due to differences in oxygen availability during the period between the end of contraction and freezing of the biopsy material. *The constant relation in normal man between phosphocreatine and lactate is of practical value since it can be used to evaluate the relation between glycolysis and the creatine kinase reaction in pathological states in which disturbances in energy metabolism can be suspected.*

There was a continued breakdown of the total adenine nucleotide pool (TAN) during the latter half of the 75 s stimulation period. TAN was decreased by a mean of 7μmol g^{-1} dry weight (Table 2). This decrease was matched by a similar increase in the IMP concentration. The mean decrease

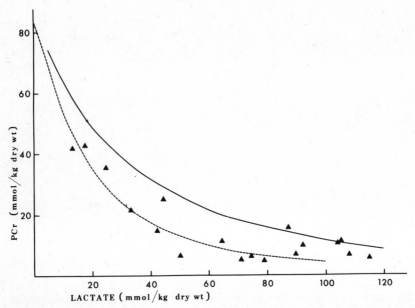

FIG. 3. Phosphocreatine and lactate in biopsy samples from the quadriceps femoris muscle during electrical stimulation at 20 Hz with occluded circulation (▲). Upper curve from voluntary contractions (Harris et al 1977). Lower curve from resting muscle during surgical knee operations. (H. Sjöholm et al, unpublished observations.)

in ATP at 75 s stimulation was 8.5 μmol g^{-1} while the AMP content was slighly increased. Total ADP was increased by about 50%. This would correspond to a pronounced increase in free ADP (about 500%), assuming that the protein-bound fraction is 90% of the ADP store at rest (Seraydarian et al 1962) and that the accumulated ADP remains unbound.

The ATP production rate was highest in the first 25 s of stimulation (Table 1, p 24) with a marked reduction when stimulation was continued. This is due to the fact that during the first 25 s both phosphocreatine splitting and glycolysis are utilized for energy production. When stimulation is continued beyond 25 s the glycolytic rate is only moderately reduced but, since the phosphocreatine store is exhausted, glycolysis is the only available energy producer and the total ATP production is markedly decreased (Table 1).

Oxidative metabolism

Biopsies taken after voluntary isometric contractions to fatigue showed phosphocreatine values in the order of 5–10% of the resting value and lactate

TABLE 2 Content of high-energy phosphates and lactate in biopsy samples from the quadriceps femoris muscle immediately after a fatiguing electrical stimulation for 75 s at 20 Hz and after 1 min recovery with intact circulation

Time after electrical stimulation (min)	ATP	ADP	AMP	TAN	IMP	ATP/ADP	Phosphocreatine	Lactate	n
0	16.05 ± 2.77 N.S.	4.22 ± 0.87 xxx	0.35 ± 0.20 x	20.62 ± 2.60 N.S.	6.01 ± 1.99 N.S.	3.97 ± 1.20 xx	9.14 ± 3.66 xxx	99.3 ± 11.3 N.S.	6
1	16.42 ± 2.31	2.44 ± 0.33	0.09 ± 0.08	18.95 ± 2.20	4.52 ± 2.15	6.87 ± 1.73	43.73 ± 2.65	87.8 ± 12.8	6
Basal*	24.0 ± 2.64	3.2 ± 0.45	0.10 ± 0.05	27.4 ± 2.52	0	7.50 ± 1.7	75.5 ± 7.63	5.1 ± 2.47	81

*Basal values from Harris et al (1974). Metabolites are expressed as mmol/kg dry muscle. Values are means ± SD. TAN, ATP+ADP+AMP. Significance of differences tested by Student's t-test. $P<0.01$, xxx; $P<0.01$, xx; $P<0.05$, x; N.S., no statistical significance.

concentrations of 80–120 μmol/g dry tissue. Repeated biopsies with arrested circulation showed unchanged values of phosphocreatine, ATP and lactate (Harris et al 1977).

When the circulation was restored, phosphocreatine was rapidly resynthesized, about 70% of the normal store being re-formed within 2 min (Harris et al 1976). It was suggested that the immediate effect of the restored circulation was to increase the availability of oxygen in the cell rather than to change cellular pH by transport of H$^+$ out of the cell. The last 30% of the phosphocreatine store was resynthesized at a lower rate after the recovery in intracellular pH. The dependence on oxygen was confirmed by incubating a tissue sample in atmospheres of oxygen and of nitrogen. Fifteen min of incubation in oxygen increased the phosphocreatine from 4% to 68% of the value at rest in spite of unchanged pH, but incubation in nitrogen left the phosphocreatine content unchanged (Sahlin et al 1979).

From these studies we suggested that the rapid resynthesis of phosphocreatine after exhaustive muscle contraction was limited by the availability of oxygen, whereas the subsequent slow phase of phosphocreatine resynthesis was limited by transport of H$^+$ out of the muscle.

This model was used to study oxidative metabolism in intact muscle. After 75 s of electrical stimulation and arrested blood flow the circulation was restored for 60 s by deflating and inflating the tourniquet and another biopsy was taken. During the 60 s circulation period there was a rapid resynthesis of phosphocreatine at a rate of 0.5 μmol g^{-1} s^{-1} with only marginal variation between the subjects (Fig. 4). This rate corresponds well with the resynthesis rate observed after voluntary contractions maintained until fatigue (Harris et al 1976). All the ATP synthesized is apparently converted to phosphocreatine, since the ATP content is unchanged and remains at a reduced level. ADP does, however, decrease so that the ATP/ADP ratio is almost back to the resting value (Table 2). *Measurement of oxidative metabolism in this way, as the rate of resynthesis of phosphocreatine after a preceding fatiguing stimulation, is proposed as a procedure for evaluating normal and pathological conditions.* Effects of training and of variation in the fibre composition of muscle could possibly be evaluated by this method in normal man, as well as mitochondrial impairment and circulatory insufficiency in pathological conditions.

Contractile properties in relation to metabolism

The tension resulting from continued electrical stimulation of the muscle under anaerobic conditions is the end result of a chain of events. Most or perhaps all of these events could undergo changes during continued stimula-

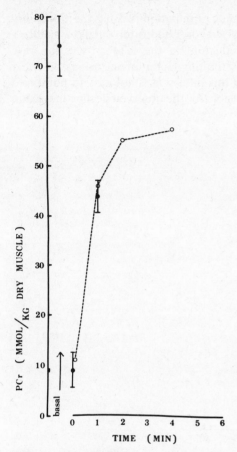

FIG. 4. The phosphocreatine content of biopsy samples taken immediately after a 75 s stimulation period and again after 60 s with intact circulation. Means ± SD. Interrupted line joins mean values of phosphocreatine content in the recovery after fatiguing isometric contractions. (Harris et al 1976.)

tion, contributing to the observed decline in tension. Excitability of the nerve and muscle membranes could change as a result of electrolyte shifts, pH changes or changes related to an altered rate of regeneration of ATP. Transmission from nerve to muscle could be altered. Conduction in the T (transverse) tubules and release of Ca^{2+} from the sarcoplasmic reticulum could be affected by ionic and pH changes. The contraction mechanism itself could be affected by a changed affinity of troponin for Ca^{2+}, due to changed pH and increased Mg^{2+} concentration when ATP is decreased. The activity of myosin ATPase could be decreased when the pH is lowered and ADP and inorganic phosphate levels are increased. The free-energy change for ATP hydrolysis is decreased during continued stimulation, raising the possibility that the energy yield per molecule ATP hydrolysed is too small to maintain cross-bridge cycling. The marked reduction in the ATP production rate during continued stimulation may be a limiting factor, giving as an end effect fewer cross-bridge attachments. With this in view it is not easy to point to a single limiting metabolic factor responsible for the observed decline in tension (Fig. 5).

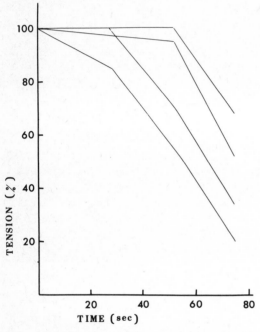

FIG. 5. Tension development in the quadriceps femoris muscle from four subjects during electrical stimulation at 20 Hz for 75 s.

Changes in relaxation rate show smaller variations between individuals in the course of continued stimulation than do the tension changes. Relaxation is also a less complex phenomenon than tension development. The rate-limiting factor for cross-bridge detachment is either the capacity for splitting ATP or the pumping back of Ca^{2+} into the sarcoplasmic reticulum. Both mechanisms depend on energy metabolism. *The relaxation rate should therefore be an indirect measure of the energy status of the muscle.*

During stimulation there was a progressive increase in the relaxation time, which was increased by about 200% after 75 s of stimulation (Fig. 6). In the recovery period when the circulation was restored there was a rapid norma-

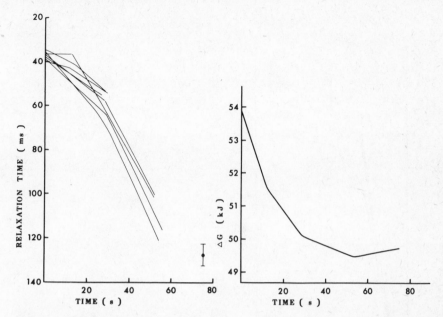

FIG. 6. *Left*. Relaxation time (measured as time taken for tension to fall from 95% to 50% of steady plateau tension) determined immediately after continued electrical stimulation at 20 Hz. Mean ± SD for six experiments with 75 s stimulation, Ī, is also shown. *Right*. Free-energy change for ATP hydrolysis during electrical stimulation at 20 Hz as calculated from metabolite contents in biopsy samples. (For calculations see Sahlin et al 1978.)

lization of the relaxation time at approximately the same rate as the resynthesis of phosphocreatine (Fig. 7).

Possible mechanisms of changes in relaxation rate

Wilkie's group (Dawson et al 1980) has recently suggested, on the basis of nuclear magnetic resonance observations in frog muscle, that the relaxation rate depends on the free-energy change for ATP hydrolysis. This seems not to be the case in human muscle because when the greatest decrease in free-energy change is observed, the relaxation time increases very little, and when there is a greater increase in relaxation time there is practically no alteration in the free-energy change (Fig. 6).

It is unlikely that lactate accumulation or pH changes as such are responsible for the changes in relaxation rate, as a rapid normalization of the relaxation rate is observed in the recovery period after stimulation with only small changes in pH or lactate content (Table 2, p 27) (Harris et al 1976). The

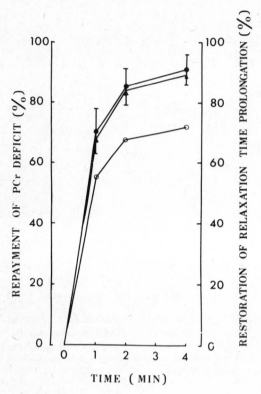

FIG. 7. Repayment of phosphocreatine deficit and restoration of relaxation time during the recovery period after muscle contraction. Lower and middle curves denote phosphocreatine repayment after fatiguing isometric contractions and exhaustive dynamic exercise respectively (data from Harris et al 1976). Upper curve: % restoration of prolonged relaxation time; means ± SD in six subjects after 75 s of electrical stimulation.

absolute level of ATP seems not to be important either, as no changes were observed during the recovery period.

ADP may be of importance, as an increased concentration (see Table 2) could both slow down cross-bridge detachment and reduce the activity of the Ca^{2+}-pumping ATPase in the sarcoplasmic reticulum membrane.

The reduced rate of ATP production when stimulation is continued coincides roughly with the slowing down of relaxation (Fig. 8). It is tempting to speculate that the ATP production rate falls below what is needed to keep the ATP/ADP ratio at a level necessary for pumping Ca^{2+} into the sarcoplasmic reticulum at a normal rate. A decrease in the ATP/ADP ratio would slow down Ca^{2+} pumping and hence relaxation. ATP/ADP ratios determined in the biopsy samples would however reflect a steady state under the prevailing

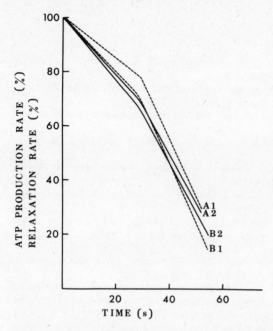

FIG. 8. Relaxation rate (A1,B1) and ATP production rate (A2,B2) for two subjects during electrical muscle stimulation. The rates are expressed as percentages, where 100% denotes initial rate and 0% denotes rate after 75 s stimulation.

conditions rather than the actual ratios during relaxation. The marked prolongation of relaxation starts when the phosphocreatine store is utilized and it is normalized when phosphocreatine is resynthesized. Thus the link between phosphocreatine and relaxation rate could be that phosphocreatine is a readily available store for rephosphorylation of ADP, so keeping ATP production at a sufficiently high level for relaxation to take place at a normal rate.

Acknowledgements

This work was supported by grants from the Swedish Medical Research Council (project 03X–2647) and the Swedish Sports Research Council (project 36/77).

REFERENCES

Bergström J 1975 Percutaneous needle biopsy of skeletal muscle in physiological and clinical research. Scand J Clin Lab Invest 35:609-616

Bergström J, Hultman E 1966 Muscle glycogen synthesis after exercise an enhancing factor localized to the muscle cells in man. Nature (Lond) 210:309-310

Bergström J, Hultman E 1967 A study of glycogen metabolism during exercise in man. Scand J Clin Lab Invest 19:218-228

Bergström J, Hultman E 1972 Nutrition for maximal sports performance. J Am Med Assoc 221:999-1006

Bergström J, Hermansen L, Hultman E, Saltin B 1967 Diet, muscle glycogen and physical performance. Acta Physiol Scand 71:140-150

Bergström J, Harris RC, Hultman E, Nordesjö L-O 1971 Energy rich phosphagens in dynamic and static work. Adv Exp Med Biol 11:341-355

Dawson MJ, Gadian DG, Wilkie DR 1980 Mechanical relaxation rate and metabolism studied in fatiguing muscle by phosphorus nuclear magnetic resonance. J Physiol (Lond) 299:465-484

Edwards RHT, Young A, Hosking GP, Jones DA 1977 Human skeletal muscle function: description of tests and normal values. Clin Sci Mol Med 52:283-290

Harris RC, Hultman E, Nordesjö L-O 1974 Glycogen, glycolytic intermediates and high-energy phosphates determined in biopsy samples of musculus quadriceps femoris of man at rest. Methods and variance of values. Scand J Clin Lab Invest 33:109-120

Harris RC, Edwards RHT, Hultman E, Nordesjö L-O, Nylind B, Sahlin K 1976 The time course of phosphorylcreatine resynthesis during recovery of the quadriceps muscle in man. Pflügers Arch Eur J Physiol 367:137-142

Harris RC, Sahlin K, Hultman E 1977 Phosphagen and lactate contents of m. quadriceps femoris of man after exercise. J Appl Physiol 43:852-857

Hultman E 1967 Studies on muscle metabolism of glycogen and active phosphate in man with special reference to exercise and diet. Scand J Clin Lab Invest 19 suppl 94:1-64

Hultman E, Bergström J 1973 Local energy-supplying substrates as limiting factors in different types of leg muscle work in normal man. In: Keul J (ed) Limiting factors of physical performance. International Symposium at Gravenbruch 1971. Georg Thieme, Stuttgart, p 113-125

Hultman E, Nilsson LH 1971 Liver glycogen in man, effect of different diets and muscular exercise. Adv Exp Med Biol 11:143-151

Hultman E, Nilsson LH 1973 Liver glycogen as a glucose-supplying source during exercise. In: Keul J (ed) Limiting factors of physical performance. International Symposium at Gravenbruch 1971. Georg Thieme, Stuttgart, p 179-189

Hultman E, Bergström J, McLennan Anderson N 1967 Breakdown and resynthesis of phosphoryl-creatine and adenosine triphosphate in connection with muscular work in man. Scand J Clin Lab Invest 19:56-66

Radner G 1962 Transactions of the Swedish Medical College of Sciences

Sahlin K, Palmskog G, Hultman E 1978 Adenine nucleotide and IMP contents of the quadriceps muscle in man after exercise. Pflügers Arch Eur J Physiol 374:193-198

Sahlin K, Harris RC, Hultman E 1979 Resynthesis of creatine phosphate in human muscle after exercise in relation to intramuscular pH and availability of oxygen. Scand J Clin Lab Invest 39:551-557

Saltin B, Karlsson J 1971 Muscle glycogen utilization during work of different intensities. Adv Exp Med Biol 11:289-299

Seraydarian K, Mommaerts WFHM, Wallner A 1962 The amount and compartmentalization of
  adenosine diphosphate in muscle. Biochim Biophys Acta 65:443-460

DISCUSSION

Edwards: This exciting paper brings us closer to seeing a relationship
between metabolic and mechanical factors. Perhaps Professor Wilkie wants
to comment on the free-energy relationships and the ATP turnover rate?

Wilkie: Yes. In Fig. 6 (p 31) you plotted relaxation time, defined as the
time to go from 95% to 50% of force during relaxation—that is, the half-time
of mechanical relaxation—against time (N.B. half-time is $0.6931 \times$ time
constant). However, what we showed (Dawson et al 1980) wasn't that the
time constant of relaxation is proportional to the free-energy change but that
the *rate constant*, which is the reciprocal of the time constant, is. Have you
looked at your results plotted in that way? If you plot the rate constant against
the free-energy change, you might get a linear relationship after all!

Sjöholm: We did that plot of relaxation rate in relation to free-energy
change and there is still not a linear relationship.

Wilkie: That is the correct basis for comparing your work and our findings
with frog muscle.

Sjöholm: Yes.

Dawson: How did you calculate the affinity (free-energy change) for ATP
hydrolysis?

Hultman: We calculated ΔG according to the following formula:

$$\Delta G = -\Delta G^0 + 5.9 . \log \left[\frac{(ADP) \, (P_i)}{(ATP)} \right]$$

Dawson: Was the concentration of ADP that you used in your calculation
the total ADP?

Hultman: Yes. We use the analytically determined ADP concentration—
that is, the total ADP.

Wilkie: One should *not* use the analytically determined (total) ADP
concentration for this purpose; it is the *free* ADP concentration that counts. In
the resting muscle that is roughly 10% of the concentration obtained by
analysis, as you said.

Newsholme: That is debateable! We are not convinced that it is as low as
you suggest. We have been trying for 10 years to measure the free ADP by a
variety of techniques but the results are extremely variable. We would not
publish them until we obtain better reproducibility between different animals.

Wilkie: In our paper we shall show two lines of evidence on this question.

One of them, valid only for frog muscle at the moment, shows that the free ADP level must be low. In the other I have re-worked some of the results of Sahlin et al (1975) on human muscle, which make sense only if the free ADP concentration is low there too.

Newsholme: The question is, how low?

Wilkie: It is as low as is demanded by the measurements of the equilibrium constant for the reaction, which is roughly 1.5 to $2 \times 10^9 \, M^{-1}$ for the reaction

$$MgADP^{1-} + PCr^{2-} + H^+ \rightarrow MgATP^{2-} + Cr^0$$
(pH range down to 7.0).

Hultman: Let us assume that 90% of the ADP value in Table 2 (p 27) is protein-bound and 10% free. How do you calculate the free ADP when the content decreases, as in the recovery phase after electrical stimulation?

Wilkie: You did indeed indicate a decrease in three items (4, 7 and 8) in Table 1 in Sahlin et al (1975). The explanation is either experimental fluctuation or the fact that in those particular experiments there was substantial breakdown of ADP to AMP and perhaps IMP. In the same table the AMP levels didn't show the expected increase; you did not analyse for IMP, so I was left in the dark. I was aware that three of your results were aberrant and I have had to leave them out of my own Fig. 4 (p 110), using an independent criterion, as just explained.

Sjöholm: We made an alternative calculation assuming 90% binding of ADP. The ΔG calculated with these assumptions further increased the discrepancy between ΔG and relaxation rate.

Wilkie: I am not so worried about that, because there may well be a difference between frogs at 4 °C and humans at higher temperatures. We haven't analysed our own results for frogs at 20 °C from this point of view yet, so we don't know whether our result is of wider significance or not.

Sjöholm: There is also a great difference in the inorganic phosphate content of frog and human muscle. The values in human muscle determined in biopsy specimens are of the order of 40 mmol/kg dry muscle (Sahlin et al 1978), compared to about 5 mmol/kg in the frog.

Wilkie: Inorganic phosphate is difficult to measure in a tissue. You have to avoid breaking down phosphocreatine and producing inorganic phosphate while freezing and extracting the sample, which gives an erroneously high value. By nuclear magnetic resonance (n.m.r.) studies in frog muscle the free internal phosphate concentration is found to be low, less than 2 mmol/kg wet weight.

Dawson: It is really very important to know whether the mechanisms of fatigue are or are not similar in frog and human muscle. We could say more about this if we treated our results on the two preparations in the same way.

Would you be willing, Professor Hultman, for us to take the results you presented here and analyse them by our own methods, as we have done with some of your published results in the past?

Hultman: Yes, of course!

Edwards: Another interesting observation was the reduced ATP turnover later in the contraction, Dr Hultman. That seems to be at variance with the n.m.r. work of Professor Wilkie and his co-workers. It is however something that we observed, too (Edwards et al 1975a). It has considerable implications for fatigue, in terms of how long the muscle can sustain force, and the relationship between energy supply and mechanical factors. The ATP turnover rate appears to be reduced in your results to about half its original value.

Dawson: How far does force decrease? What we have found, in frog muscle, is a linear relationship between *force* and the rate of ATP turnover (Dawson et al 1978), and so Dr Hultman's results are not necessarily inconsistent with our own.

Edwards: You showed a record of a contraction which was sustained for over a minute. How much force was lost in the course of any one contraction, if you related the ATP turnover rate to force at different stages of contraction?

Sjöholm: There is not a good relationship between measured force and ATP turnover rate because the variability in force between individuals is much greater than the variability in ATP turnover. This is partly because the measurement of force is related to knee extension, and the electrically stimulated thigh contains muscle groups with little or no activity as knee extensors. ATP consumption in such muscle groups will be poorly related to measured force.

Edwards: I accept that, but it is important to know whether the reduction in ATP turnover rate can be solely accounted for by the reduction in force or whether there are other reasons that might suggest an alteration in the energy economy of force maintenance.

Hultman: It is difficult to relate ATP turnover to the measured force, for the reasons given by Hans Sjöholm. A change in the contraction force measured can also be obtained by changing the position of the thigh during stimulation or in relation to the biopsy procedure. For these reasons the force measurements used are not ideal for the exact calculation of energy output.

Wiles: We find that when electrically stimulating between 20 and 40% of the quadriceps percutaneously at 20 Hz, a smooth constant-force plateau can be maintained for about 45 s before fatigue occurs. So it should be possible to say whether the ATP turnover rates are measured at constant force.

Edwards: There seems nevertheless to be a clear alteration in the energy, as indicated by heat production at different times during contraction in human muscle, suggesting that less heat is produced later on (Edwards et al

1975b). At the moment we are not clear whether that is due to an improvement in the economy of ATP utilization. Dr Jones's observations in isolated mouse muscle (Edwards et al 1975a) suggest that, but Professor Wilkie's n.m.r. studies in frog muscle do not. If we could agree that there may be an improved economy as far as heat production is concerned, can you have improved heat production without improved economy (i.e. efficiency) of ATP turnover? Or is there improved economy of ATP turnover as well and, if so, is it related to the time course of relaxation?

Hultman: If the relaxation time increases, the economy of contraction will improve.

Edwards: That is the implication, but has it been shown to be so in terms of a reduced ATP turnover for a given force?

Hultman: Not to my knowledge, and certainly not in our studies. A better measurement of the force produced will probably make such studies possible.

Newsholme: I am not clear what is meant by 'economy' in this context.

Edwards: We have various criteria for economy. We can talk about heat production per unit force, for example. In Eric Hultman's paper we are thinking of economy in terms of ATP turnover rate in relation to force at different times. Economy can also be considered in terms of the energy available from hydrolysis of ATP, as we shall hear from Professor Wilkie (p 102–114).

Newsholme: I am worried that it might be implied that all the ATP hydrolysis, at all stages, is used by the contractile system of the muscle cell. There may be other systems that utilize ATP. For example, it will be used in biochemical regulation, which is particularly important during the early stages of contraction. Furthermore, the rate of utilization will change with time. ATP will be used rapidly in the early stages but the rate might be reduced in the later stages of contraction.

Edwards: That is an interesting aspect; we haven't been thinking in those term. We await your paper with interest!

Wiles: I was surprised to see in Professor Hultman's paper that relaxation times are still markedly lengthening even after 75 s of stimulation. Our own studies in man (Wiles 1980) and those of Dawson et al (1980) in the frog tend to show that relaxation rate (expressed as a rate constant for the exponential phase of force decline) reaches a plateau after an initial decline related to the amount of contractile activity performed.

Hultman: We found only marginal changes in relaxation time after 10–15 s of stimulation; after contraction periods of 25 s and 50 s there were pronounced increases in relaxation time. In the period from 50 s to 75 s of stimulation we saw only small further increases in relaxation time. This pattern of relaxation time is very close to the change in ATP turnover rate, which decreased considerably after 25–50 s of stimulation and then stayed low during continued stimulation. If stimulation was continued beyond 75 s with an occluded circulation, very low force was obtained and no tendency to contracture was seen.

Roussos: You showed relations between the decrease in phosphocreatine and increase in lactate during contraction, and I presume you also have information on the relation between the decrease in force and in phosphocreatine. You then showed that during recovery the phosphocreatine content was rapidly restored to normal when the circulation was restored. How fast does the force come back? Is the time course of the force relationship the same as that of phosphocreatine?

Hultman: No. Phosphocreatine resynthesis is more rapid than the regeneration of force.

Roussos: What is force most closely related to during recovery—to lactate, to ATP recovery, or what?

Hultman: I have not sufficient material to answer that question. The recovery of force is also related to the stimulation frequency used, as shown by Edwards et al (1977). Low frequency fatigue remains for some hours, and is not related to ATP or lactate content. High frequency force was not measured in our studies.

Edwards: You find that force recovers more slowly than phosphocreatine; that is different from other forms of fatigue in which the action potential and force recover very rapidly. We shall hear about that later from Dr Jones (p 178–192). This emphasizes that there may be different types of fatigue in different forms of contraction.

Saltin: How large a proportion of the muscle was involved in the two types of contraction? And do you recruit fibres in a similar way when you electrically activate the muscle and when a voluntary contraction is made? In other words, are the two curves really comparable?

Hultman: In a maximum voluntary contraction all the knee extensors are utilized; that is, 100% of the muscle mass is used. Electrical stimulation of the muscle with cutaneous electrodes will activate only a part of the muscle. In our studies only 30% of the maximum voluntary contraction force is produced by the electrical stimulation. There is also a relation between force development and stimulation frequency which varies between individuals. It seems that 50 Hz produced maximum force in all subjects, while 20 Hz produced a force varying from 50 to 95% of maximum. No relation has been found between fibre type composition and percentage activation at 20 Hz. Probably all fibres are recruited during electrical stimulation, while in voluntary contraction only some of the fibres are working simultaneously.

Karlsson: We ought to consider the impact of these findings on ATP turnover and the relationship between ATP utilization and heat production during muscle contraction. An entirely different metabolic pattern might be present with electrical stimulation and with voluntary contraction. Consequently, muscle fatigue might be different in these two situations.

REFERENCES

Dawson MJ, Gadian DG, Wilkie DR 1978 Muscular fatigue investigated by phosphorus nuclear magnetic resonance. Nature (Lond) 274:861-866

Dawson MJ, Gadian DG, Wilkie DR 1980 Mechanical relaxation rate and metabolism studied in fatiguing muscle by phosphorus nuclear magnetic resonance. J Physiol (Lond) 299:465-484

Edwards RHT, Hill DK, Jones DA 1975a Metabolic changes associated with the slowing of relaxation in fatigued mouse muscle. J Physiol (Lond) 251:287-301

Edwards RHT, Hill DK, Jones DA 1975b Heat production and chemical changes during isometric contractions of the human quadriceps muscle. J Physiol (Lond) 251:303-315

Edwards RHT, Hill DK, Jones DA, Merton PA 1977 Fatigue of long duration in human skeletal muscle after exercise. J Physiol (Lond) 272:769-778

Sahlin K, Harris RC, Hultman E 1975 Creatine kinase equilibrium and lactate content compared with muscle pH in tissue samples obtained after isometric exercise. Biochem J 152:173-180

Sahlin K, Palmskog G, Hultman E 1978 Adenine nucleotide and IMP contents of the quadriceps muscle in man after exercise. Pflügers Arch Eur J Physiol 374:193-198

Wiles CM 1980 The determinants of relaxation rate of human muscle *in vivo*. PhD thesis, University of London

Muscle fibre recruitment and metabolism in prolonged exhaustive dynamic exercise

BENGT SALTIN

August Krogh Institute, Copenhagen University, Copenhagen, Denmark

Abstract The rather constant amount of glycogen found in all fibre types in human skeletal muscle provides an opportunity to study the pattern of glycogen depletion with exercise, which should give an indication of which fibres are activated to generate the force. In very light dynamic contractions repeated for hours there is a primary reliance on slow twitch (ST) fibres with no or very minor involvement of fast twitch (FT) fibres. At heavier work loads (>50% Vo_2max) ST fibres are depleted first but FT fibres begin to become depleted. Exhaustion at these work levels coincides with muscle fibres of all types being depleted of glycogen.

The crucial role of muscle glycogen in both the metabolic response to exercise and work performance is apparent. It is more difficult to explain why extramuscular substrates (plasma free fatty acids) cannot be utilized at a high enough rate to accommodate the energy turnover needed in more intense dynamic exercise. A limitation on the uptake of free fatty acids by the muscle cell rather than its transport to the cell or oxidation within it appears to be the critical factor.

In the 1960s results of studies of the utilization of glycogen in human skeletal muscle first became available. The picture emerged that glycogen breakdown during the first 20–30 minutes of a period of exercise was related to the work performed in a hyperbolic rather than linear fashion—that is, with work loads demanding more than 50–60% of the individual's maximum oxygen uptake, the utilization of glycogen was markedly enhanced (Saltin & Karlsson 1971). The question then arose as to what extent activation of skeletal muscle fibre types with different metabolic characteristics and glycogen content contributed to the observed rates of glycogen utilization. In several mammalian species the amount of glycogen is low in slow twitch (ST) fibres (Barnard et al 1971). At that time the available histochemical information also indicated differences in the glycogen levels of ST and fast twitch (FT) fibres in human

1981 Human muscle fatigue: physiological mechanisms. Pitman Medical, London (Ciba Foundation symposium 82) p 41–58

skeletal muscle (Dubowitz & Brooke 1973). The possibility then existed that at low and medium work levels ST fibres were mainly recruited, thereby explaining the moderate glycogen breakdown at these work intensities. However, more detailed quantitative determination of the glycogen content of muscle fibres revealed only very minor differences between fibre types in human skeletal muscle (Essén et al 1975). This opened up the possibility of using the glycogen depletion method to investigate the fibres and fibre types on which major reliance was placed in generating force in exercise of various intensities.

In this paper I shall summarize some of these studies. An attempt will also be made to analyse the relation between substrate availability and work performance in long-lasting dynamic efforts.

Glycogen depletion pattern in sub-maximal dynamic exercise

At very light levels (30–40% $\dot{V}_{O_2}max$) of bicycle exercise, the depletion of glycogen from individual fibres in the quadriceps muscle occurs exclusively in the ST population (Gollnick et al 1974a). At first some ST fibres stain weakly, but as exercise continues all the ST fibres become paler and some of them become depleted of glycogen. After three hours of continuous bicycle exercise many of the ST fibres are glycogen-depleted but almost all FT fibres are still completely filled with glycogen. Only a very few, if any, FT fibres are more weakly stained than an hour earlier. At this work intensity the exercise can continue for many more hours. As no systematic studies are available on glycogen depletion after three hours, we do not know whether or not FT fibres are also gradually depleted of glycogen. In a study of running 100 km, which took 9–12 hours, glycogen was still present in the exercising muscles (gastrocnemius) and most of this glycogen was found in the FT fibres (Gad & Holm 1977). It therefore appears likely that in light dynamic exercise there is an almost exclusive reliance upon ST fibres throughout the exercise period. For this to be the case the ST fibres must be able to support their metabolism by taking up substrate from the bloodstream after they have used all their stored glycogen. To what extent intramuscular lipid stores are used is still an open question.

When the oxygen demand of the exercise reaches 50–60% or more of the individual's maximum oxygen uptake, the picture is different. At first the ST fibres become empty but after a while some FT fibres also become depleted of glycogen. If the exercise continues to exhaustion, which may set in after 2–4 hours, all fibres are depleted (Gollnick et al 1974a).

The pattern of glycogen depletion that emerges at sub-maximal exercise intensities that demand close to maximum oxygen uptake (80–90% of

Vo_2max) has been reported to be of two types. In some studies ST units are gradually depleted of their glycogen and after a while some FT fibres are also depleted (Gollnick et al 1974a). This pattern is very similar to the one just described, the only difference being that the depletion is faster. In other studies, FT units have been found to be depleted before the ST fibres (Andersen & Sjøgaard 1976). This cannot be taken as a sign of a reversed recruitment order, but is rather a function of the different metabolic properties of ST and FT units. In untrained muscle the latter have a lower oxidative potential, resulting in a lower capacity both to utilize lipids and to oxidize pyruvate. Thus, when FT units are recruited, glycogen breakdown must be faster than when ST units are involved, although the total force generated by the muscle is quite low.

Some of the discrepancies just described may be attributed to the use of different exercise devices. Bicycles which are mechanically braked and have a light flywheel offer very little momentum to the exercise. Thus, in each pedal thrust a brisk contraction may be needed to keep pace. From electromyographic (EMG) studies it is known that the threshold for activation of a motor unit is reduced in brisk or ballistic contractions (Desmedt & Godaux 1977). The above description basically relates to dynamic exercise by man. In several other species the involvement of different fibre types in various gaits has been studied. It is apparent from these studies that in walking and trotting at rather low speeds, not only ST but also FT units of the fast oxidative glycolytic type (FTa) become depleted of glycogen early in the exercise (Armstrong et al 1977). In contrast, the FTb fibres do not become glycogen-depleted in these gaits—even at high speeds—but only when the animal is galloping. These findings do not detract from the concept of an orderly recruitment of motor units but do point to species differences in the properties of the fibre types that have some bearing on how they are used in various activities.

It is notable that almost all studies of glycogen depletion patterns in voluntary exercise have been evaluated by estimating the intensity of periodic acid–Schiff (PAS) staining. It appears that the intensity of staining is reasonably well related to muscle fibre glycogen content in the range of 0–80 mmol/kg wet weight of muscle (Piehl 1974) and that subjective or photometric estimations can be made accurately. At higher glycogen levels, PAS staining is saturated.

Recently, quantitative determinations of muscle fibre glycogen content in humans have been made (Essén 1978). In all essential aspects these measurements confirm the findings obtained by PAS staining. However, there is one interesting difference. The quantitative determinations have given very broad ranges for the muscle glycogen content of both fibre types at rest after a normal mixed diet (Fig. 1). This may be the real picture, or it may in part reflect the methodological difficulties in the weighing and analytical proce-

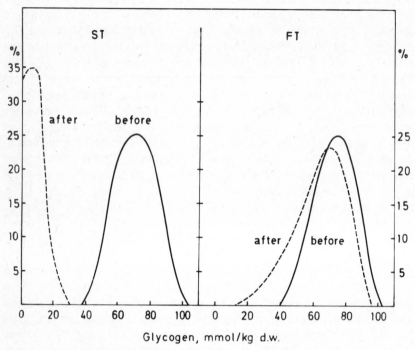

FIG. 1. Glycogen content of slow and fast twitch (ST and FT) muscle fibres from leg muscle of man (vastus lateralis) at rest (continuous line) and after one hour of exercise at close to 70% of $\dot{V}o_2$max. (D. L. Costill & B. Saltin, unpublished results.)

dures for the very small fragments used in these studies. In addition there is the problem of how much extracellular tissue adheres to individual fibre fragments when they are mechanically separated. Any one fibre fragment may include from zero to 20% of extracellular tissue.

To obtain some idea of what fraction of the voluntary contractile strength is used in dynamic exercise, the peak force for a pedal thrust during bicycle exercise was measured (Sjøgaard 1978). At work loads demanding less than the subject's maximum oxygen uptake, less than 10% of MVC is used (Fig. 2). At work loads eliciting $\dot{V}o_2$max the corresponding value was 10–15% of MVC. These measurements take into account the knee angle at which the peak force is exerted, but not the speed of contraction. When the speed of contraction is considered, the relative values are between 20 and 40% of the maximal force at a speed of contraction equivalent to 60 r.p.m. Thus, even at a work rate that exhausts a subject within some four minutes, less than half the strength of the muscle is used.

In static contractions held with the knee extensor at relative MVCs ranging from 10 to 50%, ST units are the only ones to be depleted of glycogen for

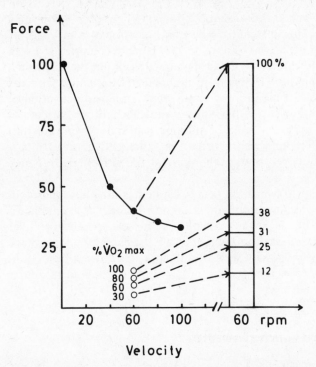

FIG. 2. The line connecting the filled circles shows the relationship between force (% MVC) (peak tension per pedal thrust) and velocity (pedal frequency) for six subjects. (Modified from Sjøgaard 1978.) It is possible that for methodological reasons the force is slightly overestimated. The tension developed per pedal thrust when exercising at 30–100% of $\dot{V}o_2$max is also shown (open circles). (Data from work by Sjøgaard 1978, Gollnick et al 1974a, 1981). To the right, these latter values are shown, expressed as a percentage of maximal force at a pedal rate of 60 r.p.m.

force development up to 15–20% of MVC (Gollnick et al 1974b). Above that level, FT units are also depleted of glycogen, which indicates that not only force but also the availability of oxygen affects the recruitment of FT units, as blood flow in the knee extensor groups with sustained contractions is not much reduced up to 15–20% of MVC.

EMG recording from active single fibres also supports the concept of an orderly recruitment of motor units in a progression from ST to FT units (Desmedt & Godaux 1977). There is very little evidence for the activation of FT motor units without ST units during high intensity exercise. There are however some reports, using either histochemical or electrophysiological techniques, indicating exceptions to the orderly recruitment of motor units described above (Secher & Nygaard 1976, Grimby & Hannerz 1973). The histochemical studies are difficult to evaluate since there is a lack of definitive proof of ST fibre involvement, because the reduction in glycogen is too small to be evident by histochemical staining. The EMG recordings from single ST and FT fibres in brisk contractions or at close to maximum voluntary strength demonstrate well that FT units may be firing moments earlier than ST units, but it is noticeable that the ST fibres are not silent (Desmedt & Godaux 1977). In a voluntary effort with synchronous preprogrammed activation of motor neurons in the motor cortex, FT units of muscle could be expected to fire before ST units, simply as a result of differences in conduction velocities and the number of interconnecting neurons. In reflex activation of motor units by input from sensory receptors in the skin, the recruitment pattern may differ.

The specific messenger to which the nervous system control responds with varying patterns of motor unit recruitment is unknown. It is easy to envisage a control system centred on the stretch receptors in muscle spindles or Golgi organs. Signals from these receptors could indicate the need for adding more motor units. In prolonged exercise, this could happen as a result of a fall in the tension-developing capacity of some motor units as their intracellular glycogen store is depleted.

Substrate availability and work performance

In dynamic exercise at work loads demanding 60–80% of $\dot{V}_{O_2}$max, the time to exhaustion in well-motivated subjects coincides with all the muscle fibres in the exercising muscles becoming depleted of glycogen. In contrast to the situation during lighter work, it appears that the exercise cannot continue at this high intensity when the glycogen is used up. This critical role of glycogen has been established in many studies. From the early experiments we know that dietary manipulation can affect the capacity for prolonged, more intense exercise (Christensen & Hansen 1939). It was shown later on that this enhanced exercise capacity is a function of the increased storage of glycogen in the exercising muscles (Bergström et al 1967). It is worth emphasizing that the short-term maximal work capacity is not improved, but rather the time to exhaustion at a given heavy sub-maximal load (i.e. a constant pace) is longer (Karlsson & Saltin 1971).

The relative role of blood-borne substrate (glucose, free fatty acids) or intramuscular triglycerides in increasing the exercise capacity is not so well documented. Muscle glycogen utilization is slightly enhanced by blocking the mobilization of free fatty acids from adipose tissue during exercise, but there is no clear-cut effect on the subject's work capacity (Bergström et al 1969). The respiratory quotient (RQ) was 0.84, indicating that the oxidation of fat was still considerable in spite of quite low levels of free fatty acids. In an attempt to evaluate this problem further we did an experiment in which the muscle glycogen content of one leg had been reduced by work on the previous day (Pernow & Saltin 1971). After exercising to exhaustion, first with one leg and then with the other, nicotinic acid being given between exercise bouts to block the availability of blood-borne free fatty acids, we found a 50% reduction in work time (from 55 to 28 min). The estimated contribution of plasma free fatty acids to the oxidation of fat was 1 g as opposed to 13 g while exercising with the second and first legs, respectively. Apparently in both situations, however, approximately 10 g of fat came from other sources (muscle and/or plasma triglycerides). Nevertheless, the reduction in availability of free fatty acids in these experiments clearly showed the importance of this source of fuel for prolonged exercise, when work is not being done at too high a fraction of the maximal capacity. In accordance with this is the finding of an improved prolonged work capacity when the level of free fatty acids in plasma is artificially raised (Rennie et al 1976, Costill et al 1977).

The extramuscular stores of glucose in man amount to 50–70 g. In prolonged exercise the exercising limbs take up glucose from the bloodstream (Wahren et al 1971). The plasma glucose concentration remains fairly constant during prolonged exercise until the subject is close to exhaustion. Thus, there is a coupling between the release of glucose from the liver and the glucose taken up by the tissues. The glucose uptake of the exercising muscles gradually increases in prolonged exercise and the uptake is higher, the heavier the work load. With limited stores of glycogen in the liver and a limited gluconeogenic capacity, however, there cannot be an unending release of glucose to match the peripheral utilization. When a certain point is reached, the plasma glucose concentration can fall to quite low values (Ahlborg et al 1974). That hypoglycaemia causes fatigue is well known (Christensen & Hansen 1939). It appears that this fatigue is more general in nature than the fatigue resulting from local muscle glycogen shortage alone (Bergström et al 1967). If glucose is given intravenously, muscle glycogen utilization is reduced slightly (Hultman 1967) and so is liver glycogen (Bagby et al 1978). When glucose is given by mouth there is also ample evidence of an improved capacity for prolonged exercise. Thus, an increase in blood-borne substrates in the form of free fatty acids and glucose has an effect on metabolism and work performance. However, none of these substrate sources appear to have as profound an effect on the metabolic response to exercise and on exercise capacity as the muscle glycogen.

This has been emphasized further in recent experiments (Gollnick et al 1981). Distinct differences in leg muscle glycogen content were obtained in subjects who had done one-legged exercise the previous day and had had a low carbohydrate diet overnight. Ordinary two-legged bicycle work was then performed at 60 and 80% Vo_2max. By means of force devices on the pedals, the two legs were made to contribute the same amount to the mechanical work output. The leg low in muscle glycogen extracted more glucose from the blood than the other leg and relied more on the oxidation of fat to compensate for the limited glycogen utilization that was possible. However, perhaps the most dramatic consequence was in the metabolism of lactate by the two legs. The leg with a normal glycogen content released lactate continuously at both work levels, whereas the leg with less glycogen took up lactate. In fact, lactate appeared to be taken up against a concentration gradient; that is, in some situations the lactate level (per litre H_2O) was slightly higher in the muscle than in the blood. This seems unlikely, since there is no known active transport of lactate in skeletal muscle. The lactate concentration in a biopsy sample, however, may not accurately reflect the concentration in all motor units. Thus, the highly oxidative ST motor units may have low lactate levels and may be primarily responsible for the uptake of lactate.

One important aspect of these results is that the adaptive responses that depend upon availability of glycogen occur locally. The behaviour of lactate is one such example, the uptake of glucose another. Glucose uptake was highest in the leg with a low glycogen content and was inversely related to the concentration of glucose 6-phosphate. Thus, factors such as the inhibition of hexokinase by glucose 6-phosphate may play a role in locally regulating glucose uptake in individual muscle fibres. In this context, it is also interesting that glucose uptake in the legs with low or normal glycogen content was closely related to the percentage of muscle fibres depleted of glycogen in each leg. Thus, it appears likely that glucose is taken up from the blood predominantly in muscle fibres low in glycogen. At the same time, we know that at work levels above 60% Vo_2max, new motor units are brought into play when initially recruited fibres are depleted of glycogen. This brings us to several interesting and important unsolved problems. One is related to the blood flow in leg muscles and its distribution within the muscle.

The available information indicates that the blood flow to the exercising muscles stays constant after its initial increase during the first minutes of the exercise, which agrees with the fact that cardiac output is essentially unchanged in prolonged exercise (Fig. 3). With a given amount of flow to the exercising muscles, but with a larger fraction of activated fibres, it is difficult to see how both glycogen-depleted and newly recruited muscle fibres can obtain an adequate share of the blood.

This leads us to other areas where our knowledge is limited. The order in which motor units are brought into work is first the ST and then the FT units. In spite of the larger involvement of FT fibres the RQ for the whole body, or RQ over the exercising legs, reveals an unchanged relative contribution of fat and carbohydrate to the energy turnover (Fig. 3).

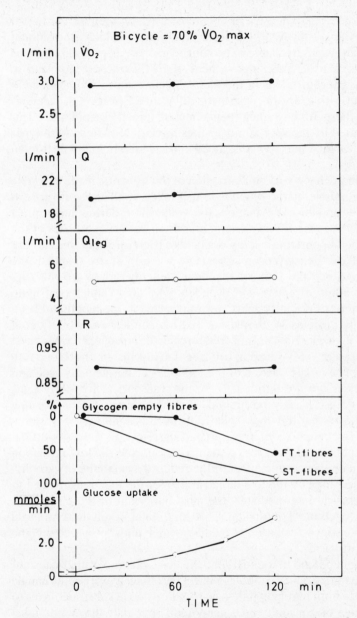

FIG. 3. Mean values of variables obtained during prolonged exercise on the bicycle at approximately 70% of V̇o₂max. The actual values are taken from different studies or estimated as follows. Oxygen uptake, RQ values and glycogen depletion (Gollnick et al 1974a); cardiac output (Saltin & Stenberg 1964); leg blood flow (one leg). Glucose uptake values are estimations based on results of Saltin et al (1976) and Henriksson (1977).

It has been noted that when FT fibres are contracting they have a higher rate of turnover of carbohydrate. It is likely, then, that the increased amount of plasma free fatty acids available in prolonged exercise is primarily utilized by the ST fibres. Further, the involvement of FT fibres does not lead to increased lactate concentrations in the blood. Lactate is probably formed in the FT fibres, but with a lack of substrate in adjacent ST fibres, this lactate is metabolized in those fibres rather than released into the capillaries. This process is facilitated by the very homogeneous mixture of various fibre types in the skeletal muscle of man. That is, an ST fibre is usually found next to an FT fibre.

Calculations show that a very small fraction of the substrate made available to the exercising muscle by the blood is in fact taken up. This is also true at exhaustion. One possible explanation for exhaustion during moderately severe exercise is the inability of skeletal muscle to extract substrates at the rates necessary to support the energy demands. Therefore, the muscles are forced to depend on their glycogen stores and, when these are emptied, the work must stop. Since the bulk of the body's energy stores are the triglycerides of the adipose tissues, one wonders why they cannot contribute more to the exercise. Free fatty acids are transported in the blood and from blood to tissue by proteins in the plasma and interstitial space (Neely et al 1972). The very high plasma levels of free fatty acids during exercise suggest that the mobilization and transport of free fatty acids in the blood are sufficient to supply a larger fraction of the total fuel than is actually used and that these processes are not limiting. If so, the transport of free fatty acids from blood into muscles may be limited. Recent observations on the dog heart *in situ* indicate that the endothelial cells of the capillaries constitute a major barrier to the uptake of free fatty acids (Rose & Goresky 1977). Further, a poor interstitial transport system contributes to the low extraction of substrates. Finally, a possible effect of the reduced availability of carbohydrate in the muscle may be the loss of substrates for the reactions needed to supply citric acid cycle intermediates, the most important of these reactions being the carboxylation of pyruvate. This loss could result in a reduced capacity of the citric acid cycle to oxidize acetyl units provided by the β-oxidation of fatty acids.

To sum up, it can be said that underlying the observation of rather constant values for leg blood flow and metabolism during prolonged exercise, dramatic changes take place both in regional flow distribution and in substrate fluxes to meet the demands of constant force generation, but that the motor units become progressively less able to develop tension, because of their limited glycogen stores and their limited capacity for utilizing plasma free fatty acids.

REFERENCES

Ahlborg G, Felig P, Hagenfeldt L et al 1974 Substrate turnover during prolonged exercise in man. Splanchnic and leg metabolism of glucose, free fatty acids, and amino acids. J Clin Invest 53:1080-1090

Andersen, P, Sjøgaard G 1976 Selective glycogen depletion in the subgroups of type II muscle fibers during intense submaximal exercise in man. Acta Physiol Scand 96:26A

Armstrong RB, Marum P, Saubert IV CW, Seeherman HJ, Taylor CR 1977 Muscle fiber activity as a function of speed and gait. J Appl Physiol Respir Environ Exercise Physiol 43:672-677

Bagby GJ, Green HJ, Katsuta S, Gollnick PD 1978 Glycogen depletion in exercising rats infused with glucose, lactate, or pyruvate. J Appl Physiol Respir Environ Exercise Physiol 45:425-429

Barnard RJ, Edgerton VR, Furukawa T, Peter JB 1971 Histochemical, biochemical, and contractile properties of red, white and intermediate fibres. Am J Physiol 220:410-414

Bergström J, Hermansen L, Hultman E, Saltin B 1967 Diet, muscle glycogen and physical performance. Acta Physiol Scand 71:140-150

Bergström J, Hultman E, Jorfeldt L, Pernow B, Wahren J 1969 Effect of nicotinic acid on physical working capacity and on metabolism of muscle glycogen in man. J Appl Physiol 26:170-176

Christensen EH, Hansen O 1939 Arbeitsfähigkeit und Ernährung. Skand Arch Physiol 81:160-175

Costill DL, Coyle E, Dalsky G, Evans W, Fink W, Hoopes D 1977 Effects of elevated plasma FFA and insulin on muscle glycogen usage during exercise. J Appl Physiol Respir Environ Exercise Physiol 43:695-699

Desmedt JE, Godaux E 1977 Ballistic contractions in man: characteristic recruitment pattern of single motor units of the tibialis anterior muscle. J Physiol (Lond) 264:673-693

Dubowitz V, Brooke MH 1973 Muscle biopsy: a modern approach. In: Major problems in neurology, vol 2. Saunders, London

Essén B 1978 Studies on the regulation of metabolism in human skeletal muscle using intermittent exercise as an experimental model. Acta Physiol Scand 102 suppl 454

Essén B, Jansson E, Henriksson J, Taylor AW, Saltin B 1975 Metabolic characteristics of fibre types in human skeletal muscles. Acta Physiol Scand 95:153-165

Gad P, Holm B 1977 Muskelfiberkomposition og enzymaktivitet samt omsætning af næringsstoffer og væske ved 100 km løb. TfL (Danish) No.5:205-213

Gollnick PD, Piehl K, Saltin B 1974a Selective glycogen depletion pattern in human skeletal muscle fibres after exercise of varying intensity and at varying pedalling rates. J Physiol (Lond) 241:45-57

Gollnick PD, Karlsson J, Piehl K, Saltin B 1974b Selective glycogen depletion in skeletal muscle fibres of man following sustained contractions. J Physiol (Lond) 241:59-67

Gollnick PD, Pernow B, Essén B, Jansson E, Saltin B 1981 Availability of glycogen and plasma FFA for substrate utilization in leg muscle of man during exercise. Clin Phys, in press

Grimby L, Hannerz J 1973 Differences in recruitment order and discharge pattern of motor units in the early and late flexion reflex components in man. Acta Physiol Scand 90:555-564

Henriksson J 1977 Training induced adaptation of skeletal muscle and metabolism during submaximal exercise. J Physiol (Lond) 270:661-675

Hultman E 1967 Studies on muscle metabolism of glycogen and active phosphate in man with special reference to exercise and diet. Scand J Clin Lab Invest 19 suppl 94:1-64

Karlsson J, Saltin B 1971 Diet, muscle glycogen and endurance performance. J Appl Physiol 31:203-206

Neely JR, Rowetto MJ, Oram JF 1972 Myocardial utilization of carbohydrate and lipids. Prog Cardiovasc Dis 15:389-396

Pernow B, Saltin B 1971 Availability of substrates and capacity for prolonged heavy exercise in man. J Appl Physiol 31:416-422

Piehl K 1974 Glycogen storage and depletion in human skeletal muscle fibres. Acta Physiol Scand suppl 402:1-33

Rennie M, Winder WM, Holloszy JO 1976 A sparing effect of increased free fatty acids on muscle glycogen content in exercising rat. Biochem J 156:647-655

Rose CP, Goresky CA 1977 Constraints on the uptake of labelled palmitate by the heart. The barriers at the capillary and sarcolemmal surfaces and the control of intracellular sequestration. Circ Res 41:534-545

Saltin B, Stenberg J 1964 Circulatory response to prolonged severe exercise. J Appl Physiol 19:833-838

Saltin B, Karlsson J 1971 Muscle glycogen utilization during work of different intensities. Adv Exp Med Biol 11:289-300

Saltin B, Nazar K, Costill DL et al 1976 The nature of the training response; peripheral and central adaptations to one-legged exercise. Acta Physiol Scand 96:289-305

Secher N, Nygaard Jensen E 1976 Glycogen depletion pattern in types I, IIA and IIB muscle fibres during maximal voluntary static and dynamic exercise. Acta Physiol Scand suppl 440:141

Sjøgaard G 1978 Force-velocity curve for bicycle work. In: Asmussen E Jorgensen K (eds) Biomechanics VI-A. University Park Press, Baltimore, p 93-99

Wahren J, Felig P, Ahlborg G, Jorfeldt L 1971 Glucose metabolism during leg exercise in man. J Clin Invest 50:2715-2725

DISCUSSION

Edwards: May we first consider whether evidence from other sources supports or refutes the pattern of recruitment that Dr Saltin has discussed?

Grimby: All the neurophysiological evidence indicates that slow twitch (ST) fibres are recruited before fast twitch (FT) fibres in motor functions where muscular fatigue can develop. There is an alternative recruitment pattern also, but this is used in quite different motor functions.

Stephens: I agree with Dr Grimby; I think there is a general rule. I myself think the rule is that motor units developing small forces are recruited first. Some of those low force units are slow twitch, but some are fast twitch units (Milner-Brown et al 1973a, Stephens & Usherwood 1977, Garnett et al 1978). In fact, we should think of recruitment order in terms of force production rather than twitch speed. There is, however, another generalization; motor units developing a lot of force are always fast twitch, but motor units developing small forces are not necessarily slow twitch. Some are fast twitch. So we must be careful about claiming that weak contractions are made with slow twitch fibres. The evidence for that is not good; weak contractions are made by slow and fast twitch motor units.

I agree that there are situations in which the order of recruitment is changed. This order of recruitment of motor units depends critically on the afferent input being generated when the movement is taking place. I have studied cutaneous afferent input, which has a powerful effect on the order of recruitment (Buller et al 1978, Stephens et al 1978, Garnett & Stephens 1978, 1981). I expect there are other examples too.

Saltin: You are implying that in light dynamic exercise some FT units will take part in the weak contractions. If so, why is there no glycogen depletion over 2–3 hours in those fibres?

Stephens: May I answer that by asking *you* why it is that in *maximal* work you appear to show ST fibres in the first few minutes with their glycogen stores preserved and the FT fibres being depleted? You cannot simply interpret glycogen depletion in terms of motor unit recruitment. It is all indirect evidence. If you want to know which motor units are involved in which tasks, you must record from them and work out their mechanical properties by a method such as spike-triggered force averaging (Milner-Brown et al 1973b)—although that method is in doubt because of the amount of motor unit synchronization in muscles (Kirkwood 1979). But if you accept the spike-triggered force averaging method, the generalization that I mentioned holds.

Saltin: If FT units are continuously activated in very light monotonous contractions repeated over an extended period of time without any glycogen breakdown, it appears that humans have a mode of biochemical regulation that is different from what is found in other species. The findings in man can be compared with the pattern of glycogen depletion in walking–trotting in the rat, guinea-pig and lion. In these species low intensity work results in glycogen depletion of fast twitch oxidative glycolytic fibres (Edgerton et al 1970, Armstrong et al 1974, 1977).

Stephens: The facts are that if we record from single motor units during weak voluntary contractions in human first dorsal interosseous muscle (1% contractions and less) and use spike-triggered force averaging to estimate twitch contraction times we find FT units active during the steady maintenance of these weak contractions. These same units can also be expected to take part in weak ballistic contractions (Desmedt & Godaux 1977 a, b, 1978).

Saltin: Perhaps some of these fast fibres may not be involved for more than the first couple of seconds, at the onset of exercise. The glycogen depletion would then be too small to be detectable.

Wilkie: What is the spike-triggered force averaging method?

Stephens: This is a method that has been used to estimate motor unit twitch characteristics in human muscle (Milner-Brown et al 1973b). An EMG needle is inserted into the muscle to record the action potentials of a single motor unit. The force output of the whole muscle is recorded. The subject is then required to maintain a steady contraction such that the unit under study fires slowly (<3 p.p.s). The unit action potential is used to trigger an averager which averages the force of contraction of the whole muscle. In this way the force of contraction generated by the single motor unit and time-locked to the unit spike is extracted by the averager from the total signal.

Bigland-Ritchie: The difference in recruitment patterns between different muscles must also be a factor in fatigue (Kukulka & Clamann 1981). Dr Stephens is presumably referring mainly to the small hand muscles, whereas Dr Saltin is talking about the large quadriceps muscle. The relative stress on motor unit types of different fatigue properties will not be the same. Dr Clamann may like to comment on this?

Clamann: I would agree with Dr Grimby and Dr Stephens that probably the units are recruited in fatiguing contractions in order of increasing size, in accordance with the Size Principle (Henneman et al 1965, 1974). On Dr Bigland-Ritchie's point, it has been my experience in a small hand muscle that all the units are recruited by about 40% of maximum voluntary contraction, whereas in a larger muscle like the biceps, units are still being recruited at 80–90% of MVC (Kukulka & Clamann 1978, 1981). The total force of contraction depends both on the discharge rates of the active motor units and on the number recruited, and the relative significance of rate coding and recruitment depend in turn on the particular muscle studied. This will of course affect the fatigue properties of that muscle.

I wanted to ask a different question. Dr Saltin was talking about glycogen depletion, which ocurs very early in a fatiguing contraction. If you could drive a slow unit and a fast unit under identical conditions, is it possible that the glycogen depletion of the slow unit would require a longer period of time? Is that why you might see glycogen in fast units depleted more quickly?

Saltin: Yes, you are right. It is worth emphasizing that in the type of work discussed the force generated by the muscle in each contraction is a small fraction of their maximal strength. Even at the highest work load, less than 40% of the strength was used, if we take the speed of contraction into consideration (Sjøgaard 1977). Second, a sign of the importance of the generation of force and speed of contraction for the involvement of FT units may come from the observation that a brisk contraction may lead to an 'earlier' involvement, which is then accompanied by rather rapid depletion of glycogen in these units.

Edwards: We would expect that, from the relative metabolic rates of the two fibre types. May we now consider the force generated by these units? There is evidence to support the same orderly recruitment pattern from measurements of relaxation rate in humans (Wiles et al 1979).

Wiles: Relaxation rate measurements in voluntary contractions in the quadriceps muscle indicate that the muscle fibres (presumably Type I, ST) used at low forces (e.g. 10% maximum) are slower than the average of those used in maximum contractions. We obtain a ratio of 2:1 for the estimated relaxation rates of Type II and Type I fibres (Wiles et al 1979). It has proved difficult to reproduce these results in the adductor pollicis, possibly because of differences in the recruitment pattern or in the fibre type composition, or both.

Edwards: Another question is whether the intrinsic force per unit cross-sectional area of the motor units is the same in the two fibre types. This has considerable implications, as John Stephens mentioned. It also has implications for muscle disease, where atrophy of the glycolytic (Type II) fibres is a common finding.

Saltin: I know of no information on force in different muscle fibre types in man. However, as indirect evidence, it is surprising how similar maximum voluntary contractions (MVCs) can be in people with low or high percentages of ST fibres. Thorstensson & Karlsson (1976) and Thorstensson et al (1977) have suggested that a muscle with many FT fibres is stronger (has a higher MVC) than a muscle with many ST fibres, but so long as no information on total cross-sectional area is available, no conclusions can be drawn. In an unpublished study from our laboratory (E. Nygaard, Y. Suzuki & K. Jørgensen) the force–velocity curve was established for the elbow flexors. With computerized tomography and muscle fibre typing the area of ST and FT fibres in the biceps could be determined. MVC was primarily related to total cross-sectional area and not to FT fibre area.

Edwards: Would you like to enlarge on the factors governing the local circulation of the muscle? You postulated an axon reflex, but does one need to postulate any neurogenic mechanism if, as some people think, the stimulus for dilatation in muscle vessels is the level of phosphate or potassium or ATP, which are direct consequences of muscle fibre activity?

Saltin: In prolonged exercise (2–4 hours) the EMG of exercising muscle gradually increases, suggesting that more motor units are recruited to generate the force needed to do the work. Information on glycogen depletion also suggests that new muscle fibres are gradually brought into play. If more and more of the muscle is activated, where is the blood flow distributed in the muscle? The question is pertinent, since the available data indicate that leg blood flow ($\approx$muscle blood flow) remains constant in prolonged exercise (Wahren et al 1971, Saltin et al 1976). The question is now: is the blood going to the fibres which were initially activated or to the newly recruited fibres?

Our concept of open and closed capillaries may have to be changed, at least partly. Direct measurements of flow in individual capillaries at rest and during contraction suggest that the speed of flow may vary greatly between neighbouring capillaries. In some it is close to zero, in others quite high, but there is the same flow at rest in all capillaries (Eriksson & Myrhage 1972).

The number of erythrocytes in the capillaries varies also, in relation to the diameter of the terminal arterioles (Klitzman & Duling 1979). If the diameter is small, the orifice of the capillary is small and few or no erythrocytes can enter, while capillaries coming from a dilated arteriole are filled with erythrocytes.

Could it be that early in exercise the capillaries around contracting fibres have a high rate of perfusion with both plasma and erythrocytes? Later, when more muscle fibres are activated, the capillaries of the newly recruited fibres have relatively faster flow and more erythrocytes than the glycogen-empty muscle fibres—the idea being that the latter fibres depend to a large extent on substrates (free fatty acids and glucose) taken up from the bloodstream. Such a regulation of flow within the muscle may be served by mechanisms that are not yet understood (cf. Honig 1979, Honig & Gayeski 1980).

Donald: Is there any reason why the blood flow should not be the limiting factor? The whole exercising muscle could remain vasodilated and the capillaries of all the fibres fed fully. Your idea of the shutting down of the capillaries of fibres that are likely to be depleted of glycogen is essentially segregating exhausted muscle, which strikes me as generally undesirable.

Edwards: Could this not be explained if the maintenance of vasodilatation depends on the constant flux of the hypothetical stimulating factors? As soon as those muscle cells have ceased to generate force, for reasons to do with glycogen depletion or excitation–contraction coupling failure, or problems in membrane excitation, they would also cease to generate such factors. It is equally possible then that the remaining blood will automatically go in the direction of the newly recruited fibres.

Hermansen: What is the evidence that Type I (ST) fibres do not take part in the contraction? Perhaps they take part, but at a slower rate of tension production, and the uptake of glucose keeps them exercising at a low intensity. So they produce tension but less than when they have all the glycogen needed for fuel. So contraction is regulated, or geared, by the amount of fuel that the blood supplies.

Clamann: What levels of force are you talking about, Professor Saltin? Type II (fast twitch) fibres aren't very well supplied with blood vessels.

Saltin: Type II fibres in man are surprisingly well supplied with capillaries. There isn't much difference between the two kinds of fibre types in man in this respect, partly because FT fibres are frequently found next to ST fibres and they share capillaries.

What Dr Hermansen says could well be true but, if you argue along that line, there will be a discrepancy in the amount of flow available to the newly recruited fibres because of demand for substrate in the substrate-depleted fibres. But the depleted fibres don't need much oxygen because they can only take up a certain, rather small amount of substrate. Therefore they receive more oxygen than they need, unless there is anaemia or some other defect of oxygen transport.

Donald: If people exercise quite severely and you then stop them exercising and keep them standing still, they fall down, because their muscles remain vasodilated for some time—15 minutes or more—but the cardiac output falls. So I am doubtful about such a subtle readjustment of parts of the capillary bed in response to these biochemical events.

Edwards: Isn't the reason why people fall over after extreme exercise related to the capacitance vessels? They may be pooling blood in dilated capacitance vessels that have enlarged to cope with another conflicting demand in long-continued exercise, namely temperature regulation.

Donald: I accept that.

REFERENCES

Armstrong RB, Saubert CW, Sembrowich WL, Shepherd RE, Gollnick PD 1974 Glycogen depletion in rat skeletal muscle fibres at different intensities and durations of exercise. Pflügers Arch Eur J Physiol 352:243-256

Armstrong RB, Marum P, Saubert CW, Seeherman HJ, Taylor CR 1977 Muscle fiber activity as a function of speed and gait. J Appl Physiol 43:672-677

Buller NP, Garnett R, Stephens JA 1978 The use of skin stimulation to produce reversal of motor unit recruitment order during voluntary muscle contraction in man. J Physiol (Lond) 277:1P-2P

Desmedt JE, Godaux E 1977a Ballistic contractions in man: characteristic recruitment pattern of single motor units of the tibialis anterior muscle. J Physiol (Lond) 264:673-693

Desmedt JE, Godaux E 1977b Fast motor units are not preferentially activated in rapid voluntary contractions in man. Nature (Lond) 267:717-719

Desmedt JE, Godaux E 1978 Ballistic contractions in fast and slow human muscles: discharge patterns of single motor units. J Physiol (Lond) 258:185-196

Edgerton VR, Simpson DR, Barnard RJ, Peter JB 1970 Phosphorylase activity in acutely exercised muscles. Nature (Lond) 225:866-867

Eriksson E, Myrhage R 1972 Microvascular dimensions and blood flow in skeletal muscle. Acta Physiol Scand 86:211-222

Garnett R, Stephens JA 1978 Changes in the recruitment threshold of motor units in human first dorsal interosseous muscle produced by skin stimulation. J Physiol (Lond) 282:13P-14P

Garnett R, Stephens JA 1981 Changes in the recruitment threshold of motor units produced by cutaneous stimulation in man. J Physiol (Lond) 311:463-473

Garnett RAF, O'Donovan MJ, Stephens JA, Taylor A 1978 Motor unit organization of human medial gastrocnemius. J Physiol (Lond) 287:33-43

Henneman E, Somjen G, Carpenter DO 1965 Functional significance of cell size in spinal motoneurons. J. Neurophysiol 28:560-580

Henneman E, Clamann HP, Gillies JD, Skinner RD 1974 Rank order of motoneurons within a pool: Law of Combination. J Neurophysiol 37:1338-1349

Honig CR 1979 Contributions of nerves and metabolites to exercise vasodilation: a unifying hypothesis. Am J Physiol 236:H705-H719

Honig CR, Gayeski TEJ 1980 Capillary recruitment and de-recruitment in exercise; relation to control mechanisms and tissue pO_2. Proc Int Union Physiol Sci 14:23 (abstr)

Kirkwood PA 1979 On the use and interpretation of cross-correlation measurements in the mammalian central nervous system. J Neurosci Methods 1:107-132

Klitzman B, Duling BR 1979 Microvascular hematocrit and red cell flow in resting and contracting striated muscle. Am J Physiol 237:H481-H490

Kukulka CG, Clamann HP 1978 Recruitment and discharge properties of human motor units in low to high force isometric contractions. Neurosci Abstr 4:940

Kukulka CG, Clamann HP 1981 Comparison of the recruitment and discharge properties of motor units in human branchial biceps and adductor pollicis during isometric contractions. Brain Res, in press

Milner-Brown HS, Stein RB, Yemm R 1973a The orderly recruitment of human motor units during voluntary isometric contractions. J Physiol (Lond) 230:359-370

Milner-Brown HS, Stein RB, Yemm R 1973b The contractile properties of human motor units during voluntary isometric contractions. J Physiol (Lond) 228:285-306

Saltin B, Nazar K, Costill DL et al 1976 The nature of the training response; peripheral and central adaptations to one-legged exercise. Acta Physiol Scand 96:289-305

Sjøgaard G 1977 Force–velocity curve for bicycle work. In: Asmussen E, Jorgensen K (eds) Biomechanics VI-A. University Park Press, Baltimore, p. 93-99

Stephens JA, Usherwood TP 1977 The mechanical properties of human motor units with special reference to their fatigability and recruitment threshold. Brain Res 125:91-97

Stephens JA, Garnett R, Buller NP 1978 Reversal of recruitment order of single motor units produced by cutaneous stimulation during voluntary muscle contraction in man. Nature (Lond) 272:362-364

Thorstensson A, Karlsson J 1976 Fatiguability and fibre composition of human skeletal muscle. Acta Physiol Scand 98:318-322

Thorstensson A, Larson L, Tesch P, Karlsson J 1977 Muscle strength and fiber composition in athletes and sedentary man. Med Sci Sports 9:26-30

Wahren J, Felig P, Ahlborg G, Jorfeldt L 1971 Glucose metabolism during leg exercise in man. J Clin Invest: 50:2715-2725

Wiles CM, Young A, Jones DA, Edwards RHT 1979 Relaxation rate of constituent muscle-fibre types in human quadriceps. Clin Sci (Oxf) 56:47-52

Relevance of muscle fibre type to fatigue in short intense and prolonged exercise in man

JAN KARLSSON, BERTIL SJÖDIN*, IRA JACOBS and PETER KAISER

Laboratory for Human Performance, Department of Clinical Physiology, Karolinska Hospital, 104 01 Stockholm, Sweden

Abstract It has been suggested that the histological and histochemical features of human muscle are important in determining performance capacity. The relationship between muscle fibre types (Type I, slow twitch fibres; Type II, fast twitch fibres) and performance on standardized tests has been studied in subjects accustomed to physical exercise, and related to their patterns of lactate metabolism, expressed as the onset of blood lactate accumulation (OBLA). This variable was found to be the best predictor of endurance capacity of the variables studied. It is suggested that in healthy male subjects muscle lactate is crucial in short, intense forms of exercise (the higher the lactate formation, the better the performance) and also in prolonged, 'endurance' forms of exercise (the later the onset of lactate formation, the higher the sustainable exercise intensity). In subjects with a high proportion of fast twitch fibres, more lactate will be formed at the same exercise intensity. This is advantageous for short intense exercise but impairs endurance performance. The deleterious effects induced by glycogen depletion were studied and found to be most pronounced in subjects rich in fast twitch (glycogen-dependent) fibres. Indications were also obtained that muscular performance is regulated in different ways in males and females. In women an inverse relationship was found between fast twitch fibres and muscle power, and between fatigue and lactate concentration, whereas direct relations were found in men.

In many physiology textbooks it is stated that neuromotor control and muscle contractility are prerequisites for 'strength-type' performance, whereas well-developed central circulatory functions are necessary and perhaps of critical importance for 'endurance-type' exercise. Almost a decade ago the first

Present address: National Defense Research Institute, S-105 01 Stockholm, Sweden.

1981 Human muscle fatigue: physiological mechanisms. Pitman Medical, London (Ciba Foundation symposium 82) p 59-74

observations on skeletal muscle fibre types in athletes were published indicating a high proportion of Type I or slow twitch (ST) muscle fibres in long-distance and marathon runners, while weight lifters, jumpers, shotputters and so on tended to have more Type II or fast twitch (FT) muscle fibres than untrained subjects (cf. Karlsson 1980). Although these observations were in line with earlier observations in animal experiments on the significance of the histological and histochemical properties of muscle for performance, they were regarded as new and focused the interest of many exercise physiologists on the impact of the periphery on the control of the central circulation in man. Our research group has concentrated on the study of the relationship between muscle fibre types and standardized muscle performance, such as single maximal contractions or a series of consecutive contractions. The approach is based on a concept developed by A. V. Hill (1965) and applied by other groups with success to heart muscle (Braunwald 1976). In some experiments we also studied the significance of muscle fibre distribution in more conventional exercise tests, such as treadmill or bicycle ergometer exercise.

Subjects and methods

Unless otherwise stated, the female and male subjects studied were accustomed to physical exercise, although not athletes. We examined the muscle fibre types in needle biopsy specimens obtained from the lateral portion of the thigh muscle (vastus lateralis muscle), analysing them as described by Tesch (1980). In some studies specimens were taken after exercise for analysis of the lactate and glycogen content (Tesch 1980). In addition, in other studies, blood samples were drawn for determination of lactate concentrations by a modified version of the method described by Barker and Summerson (see Ström 1949). Isokinetic knee extensions were used as our exercise task for assessing maximal muscle force generation at different speeds of contraction, as well as muscle power and force loss in fatigue experiments, using a model of repeated contractions introduced by Thorstensson (1976). In addition, muscle power has been evaluated by the Wingate test (Bar-Or et al 1980). Different measures of endurance capacity, such as the exercise load corresponding to the onset of blood lactate accumulation (OBLA), oxygen uptake at a steady state, as well as maximal oxygen uptake ($\dot{V}o_2$max), have been determined during treadmill running as described by Sjödin et al (1980).

Results and discussion

Short, intensive exercise compared to endurance exercise

Thorstensson (1976) showed in males that having a high proportion of FT muscle fibres was of only borderline significance for force production. The great advantage of this fibre pool was in relation to repeated maximal muscle contractions—that is, to muscle power (Fig. 1). Bar-Or et al (1980) showed that this was also true for short, intense bicycle ergometer exercise as indicated by performance in the Wingate test (Fig. 1). Tesch (1980),

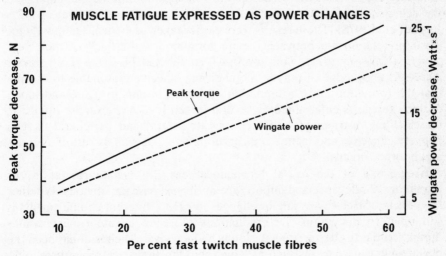

FIG. 1. Muscle fatigue expressed as peak torque decrease in the Thorstensson test (Thorstensson 1976) and power decrease in the Wingate test (Bar-Or et al 1980). Both test procedures measure muscle power generating capacity during short-term, maximal exercise in the form of a one-leg test (Thorstensson test) or a two-legged test (Wingate test).

continuing the line of Karlsson (1971) and Sjödin (1976), found that the metabolic profile and glycogenolytic activity of the FT muscle fibre population in males were of central importance in terms of both muscle power generating capacity as well as the endurance capacity of the exercising muscle. Whereas a high potential for lactate formation was synonymous with a large 'anaerobic energy output', it impaired muscle contractility and was consequently a limiting factor for sustained muscle activity. In field studies in males this concept was further developed in relation to running performance over different distances (O. Inbar, personal communication 1980). Maximal running speed over 40 m was positively correlated to the percentage of FT muscle fibres ($r = 0.73$, $n = 21$), whereas running performance over 2000 m showed the opposite pattern—a high correlation between a large proportion of ST fibres and speed ($r = 0.60$, $n = 21$). Earlier unpublished results obtained from a similar population of subjects indicate that over distances between 1000 and 2000 m there are low correlations between measures of endurance capacity such as maximal oxygen uptake, and running performance.

Lactate metabolism in relation to muscle fibre types

Earlier studies have shown a positive relationship between the accumulation of lactate in active muscle tissues and in venous blood on the one hand, and the delivery capacity for molecular oxygen on the other (Saltin et al 1971, Jorfeldt et al 1978). These results were earlier taken as evidence in muscle for a causal relationship between lactate formation and a lack of molecular oxygen (Karlsson 1971). This concept is now debatable, as a result of our increased knowledge of muscle metabolism. It seems reasonable to assume that the formation and accumulation of lactate in muscle tissue and blood during exercise is determined by several features, such as exercise intensity, muscle fibre distribution and recruitment, central and peripheral oxygen delivery capacity, and lactate utilization in active as well as resting muscles and in other organs.

Irrespective of the causal biochemical basis of lactate formation, its physiological relevance and importance during exercise are obvious. Whether the effects of lactate are due to changes in pH in different compartments is still open to discussion. Determinations of blood lactate concentrations during standardized exercise have lately had a revival. The result has been the development of concepts such as 'anaerobic threshold', 'aerobic threshold', and 'lactate breaking point'. In our laboratory we refer to OBLA (the *o*nset of *b*lood *l*actate *a*ccumulation) and use the exercise intensity corresponding to a blood lactate concentration of 4 mmol l^{-1} as a measure of endurance exercise capacity (Fig. 2). Under the prevailing exercise test conditions the blood lactate concentration can be considered as an integrated measure of the actual lactate metabolic status.

We have found that this variable for lactate metabolism, expressed as the speed corresponding to OBLA (V_{OBLA}), is a better predictor of endurance running capacity than any other performance variable studied, including maximal oxygen uptake (Fig. 3). V_{OBLA} is correlated to variables such as the muscle fibre composition, maximal oxygen uptake, running technique, and muscle fatigue indices (B. Sjödin & I. Jacobs, unpublished observations 1980). It seems reasonable to propose that in healthy male subjects, accustomed to physical exercise and training, muscle lactate metabolism is crucial not only in short, intense exercise (the higher the lactate formation, the better

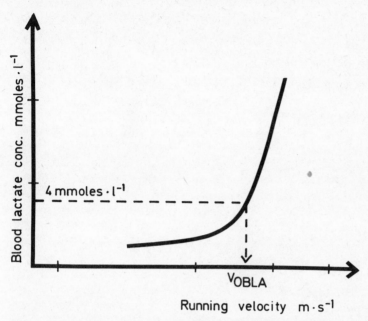

FIG. 2. Relationship between blood lactate concentration and exercise intensity during a typical OBLA test (*onset of blood lactate accumulation*). The exercise intensity is increased every 4th minute in such a way that the oxygen deficit for each stepwise increase in intensity is as small as possible. The exercise intensity corresponding to a blood lactate concentration of 4 mmol l^{-1} is defined as the subject's V_{OBLA} for treadmill running or W_{OBLA} for cycle ergometer exercise.

the performance) but also in endurance exercise (the later the onset of lactate formation, the higher the sustainable running speed). In general it can be concluded that the more FT muscle fibres are recruited, the more lactate is formed and accumulated. As a consequence a subject with a high proportion of FT muscle fibres will form more lactate at the same absolute or relative exercise intensity than one with a lower proportion. It seems reasonable to suggest that these aspects of lactate metabolism are of prime physiological relevance to muscle power generating capacity and significantly contribute to endurance capacity.

Indications of sex differences

It has been emphasized that the results presented here so far have been obtained in males used to physical exercise. The obvious question is to what extent these observations are relevant to females. Recent studies on the exercise capacity of women, based on modern exercise physiology concepts,

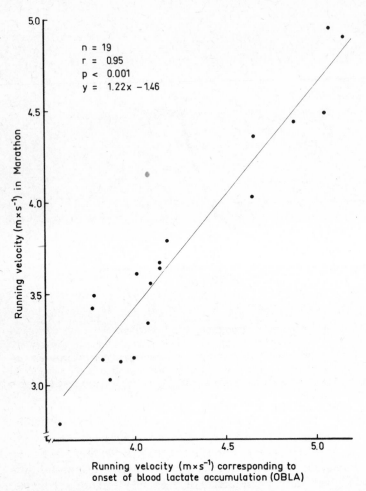

FIG. 3. Relationship between average running speed for 18 male subjects participating in the Stockholm Marathon, 1979, and V_{OBLA} (see Fig. 2).

are scarce. Komi & Karlsson (1979) presented results that indicated not only differences in absolute values of muscle strength, muscle power, maximal oxygen uptake, and so on, but even different control mechanisms in the two sexes. Thus, in males there is a positive correlation between the percentage of FT muscle fibres and muscle power, according to a number of studies; but in females, Komi & Karlsson (1979) found the opposite, namely a negative relationship. Jacobs & Tesch (1980) have recently confirmed this in females (Fig. 4). They also found a negative correlation between fatigue and muscle

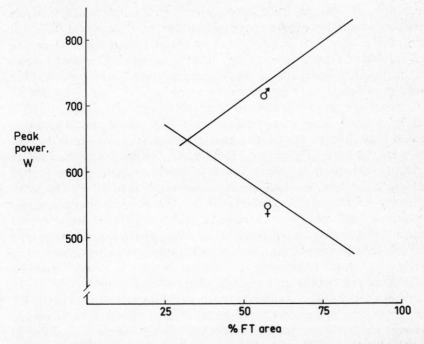

FIG. 4. Relationship for both sexes between peak muscle power generated during short-term, maximal-intensity exercise and muscle fibre type composition, expressed as the relative area occupied by the fast twitch (FT) muscle fibres. (Adapted from Jacobs & Tesch 1980.)

lactate concentrations after exercise, whereas the opposite has been shown in males by Tesch (1980).

At this stage the observations are too scarce and too fragmentary to be the basis for further speculations. It seems reasonable, though, to emphasize that the general application of observations made in studies using male subjects may be restricted.

Glycogen exhaustion and muscle fibre types

It is a well-established fact that, providing the exercise intensity is high enough (>60–70% $\dot{V}o_2$max), the local content of the main precursor for muscle glycogenolysis—muscle glycogen—is one of the factors limiting exercise capacity.

It seems reasonable to question to what extent the muscle fibre composition of individual subjects is of significance for this fatigue phenomenon. Juhlin-Dannfelt et al (1977) showed that the rate of glycogen depletion at the same relative exercise intensity was higher in subjects richer in FT muscle fibres. They would thus experience fatigue induced by glycogen exhaustion at an earlier stage than their counterparts with a large percentage of ST muscle fibres. It could be expected, in addition, that the subjects with more FT fibres would be impaired differently, as they depend more on their FT muscle fibre population, which is more glycogenolytic in its metabolic profile, while ST muscle fibres are more oxidative in their metabolism (cf. Karlsson 1980). Jacobs et al (1980) have studied the effect of generalized muscle glycogen exhaustion, as opposed to selective glycogen exhaustion, in which only ST fibres were depleted of glycogen. The former condition was obtained in laboratory experiments in which a variety of exercise regimens were applied to achieve as general and total glycogen depletion as possible. The latter condition was obtained by marathon running. The different glycogen depletion patterns were confirmed by periodic acid–Schiff (PAS) histochemical staining of muscle samples (cf. Piehl 1974). After selective depletion of only the ST muscle fibre population, no significant impairment was observed in either muscle force (single knee extensions) or muscle power (repeated maximal knee extensions). But when in addition the FT muscle fibres were depleted of their glycogen, a special pattern was observed: subjects rich in FT muscle fibres showed a greater impairment, whereas those rich in ST muscle fibres were negligibly affected and maintained normal muscle contractility in terms of strength (Fig. 5).

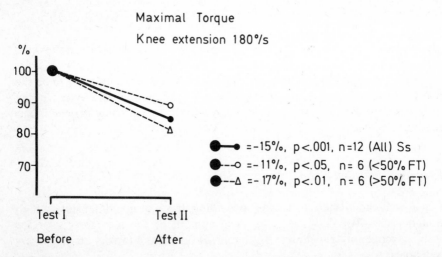

FIG. 5. The reduction in maximal torque produced during knee extension at an angular velocity of $180°$ s^{-1} after exercise-induced glycogen exhaustion of both fast twitch (FT) and slow twitch (ST) muscle fibres. The test II value (after glycogen exhaustion) is expressed relative to the value before glycogen exhaustion (I) (From Jacobs et al 1980.)

This observation further illustrates the significance of muscle fibre composition for endurance type exercise. A small percentage of FT fibres means a low rate of lactate formation and therefore only mild or no impairment due to accumulation of lactate after exercise. At the same time, a small percentage of FT fibres means a moderate expenditure of glycogen during intense exercise and the ability to depend on other substrates when glycogen exhaustion is induced.

β-Blockade and muscle fibre types

We mentioned earlier that lactate metabolism during exercise in healthy male subjects depends, among other things, on exercise intensity, muscle fibre distribution and recruitment, and central and peripheral circulatory capacities. It is also obvious that the ST muscle fibre population, depending on oxidative metabolism, will depend more on an effective regulation of the central circulation than the FT fibres. It has been suggested that there are nervous loops which mediate information from and about the periphery and consequently allow a more rapid adaptation and regulation of the circulation (cf. Rowell 1974). In trained males, positive relationships have been shown between the percentage of ST muscle fibres and maximal oxygen uptake, blood pressure and leg blood flow (cf. Karlsson 1980). This might indicate that a mechanism similar to that suggested above may be present and be related to the different muscle fibre types in man. Irrespective of these speculations, we were interested in the possibility of experimentally interfering with the central circulation and studying the effect on muscle performance. After acute β-adrenergic receptor blockade, male subjects rich in ST muscle fibres were more affected in terms of decreases in heart rate and blood pressure responses during submaximal bicycle exercise than subjects rich in FT muscle fibres (P. Kaiser, S. Rössner, L. Kaijser & J. Karlsson, unpublished observations 1980). They were also more impaired in running 2000 m than those rich in FT muscle fibres. In none of the subjects was muscle force as such affected by the β-blocking drug. Yet during two-legged, short, intense exercise (the Wingate test) subjects rich in ST muscle fibres were again more impaired. This pattern of reaction to drug treatment persisted irrespective of whether an unselective (atenolol) or a selective (propranolol) β-blocker was given. The only exception was the running test, in which a 20% impairment was evident with propranolol but less than 10% with atenolol (Fig. 6). It is still too soon in our experimental programme for us to draw any conclusions about differential reactions related to muscle fibre type distribution. It is tempting, however, to speculate about possible different regulatory mechanisms and whether these may include, in addition to 'central' reaction

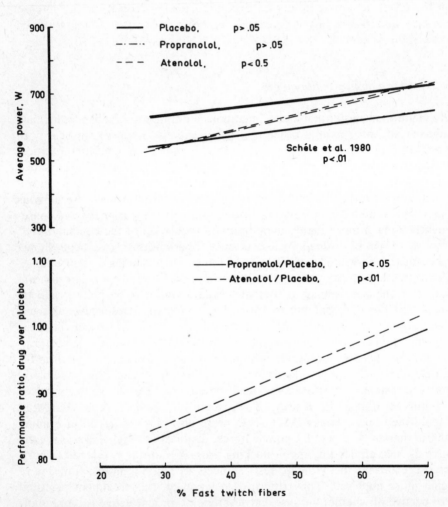

FIG. 6. Running performance (2000 m) after receiving a placebo and after acute β-adrenergic blockade with propranolol (80 mg) and atenolol (100 mg) in a double-blind study. The upper panel depicts the absolute values and the lower panel the ratio of the values in drug-treated subjects and those given placebo.

patterns, a relatively peripheral reaction pattern—for example, in the ability to form lactate in the muscles.

Summary

Muscle fibre distribution pattern is an important feature of physical performance in trained male subjects. The higher the percentage of fast twitch (Type II) muscle fibres, the higher the potential for lactate formation in the recruited musculature. This is an advantage in short, intense exercise but impairs endurance performance. Indicators of lactate metabolism under steady-state conditions are better predictors of endurance performance than maximal oxygen uptake.

We have found indications for alternative mechanisms for regulating muscle performance in females. The most obvious difference between the sexes was an inverse relation in women between the percentage of FT muscle fibres and muscle power production, as well as between muscle fatigue and lactate concentrations, in contrast to men, where direct relations are found.

The acute effects of two different forms of intervention in lactate formation have been studied: the effect of glycogen exhaustion, and the effect of reduced central circulatory capacity on muscle performance. The effect of glycogen loss was most pronounced in subjects rich in FT (glycogen-dependent) muscle fibres whereas reduced circulatory capacity, and consequently reduced oxygen delivery, affected subjects with muscles rich in ST fibres to a greater extent. It seems reasonable to suggest that any explanation of the observations must be based to a major extent on the different metabolic properties of the two main muscle fibre types in human skeletal muscle.

REFERENCES

Bar-Or O, Dotan R, Inbar O, Rothstein A, Karlsson J, Tesch P 1980 Anaerobic capacity and muscle fibre type distribution in man. Int J Sports Med 1:82-85

Braunwald E 1976 Protection of the ischemic myocardium. Introductory remarks. Circulation 53 (3 suppl 1):1-2

Hill AV 1965 Trails and trials in physiology. Edward Arnold, London

Jacobs I, Kaiser P, Tesch P 1980 Muscle strength and fatigue after selective glycogen depletion in human skeletal muscle fibers. Eur J Appl Physiol, in press

Jacobs I, Tesch P 1980 Short time, maximal muscular performance: relation to muscle lactate and fiber type in females. In: Venerondo A (ed) Proc Int Congr Women and Sport, Rome, in press

Jorfeldt L, Juhlin-Dannfelt A, Karlsson J 1978 Lactate release in relation to tissue lactate in human skeletal muscle during exercise. J Appl Physiol 44:350-352

Juhlin-Dannfelt A, Jorfeldt L, Hagenfeldt L, Hultén B 1977 Influence of ethanol on non-esterified fatty acid and carbohydrate metabolism during exercise in man. Clin Sci Mol Med 53:205-214

Karlsson J 1971 Lactate and phosphagen concentrations in working muscle of man with special reference to oxygen deficit at the onset of work. Acta Physiol Scand suppl 358

Karlsson J 1980 Localized muscular fatigue: metabolism and substrate depletion. Exercise Sport Sci Rev 7:1-42

Komi PV, Karlsson J 1979 Physical performance, skeletal muscle enzyme activities, and fibre types in monozygous and dizygous twins of both sexes. Acta Physiol Scand suppl 462:1-28

Piehl K 1974 Glycogen storage and depletion in human skeletal muscle fibres. Acta Physiol Scand suppl 402:1-33

Rowell LB 1974 Human cardiovascular adjustments to exercise and thermal stress. Physiol Rev 54:75-159

Saltin B, Gollnick P, Piehl K, Eriksson B 1971 Metabolic and circulatory adjustments at onset of exercise. In: Gilbert A, Guille P (eds) Onset of exercise. University of Toulouse Press, Toulouse, p 63-76

Sjödin B 1976 Lactate dehydrogenase in human skeletal muscle. Acta Physiol Scand suppl 436:5-32

Sjödin B, Linnarsson D, Wallensten R, Schéle R, Karlsson J 1980 The physiological background of onset of blood lactate accumulation (OBLA). In: Komi PV (ed) Proc Int Symp Sports Biology, Vierumaki, Finland. University Park Press, Baltimore, in press

Ström G 1949 The influence of anoxia on lactate utilization in man after prolonged muscular work. Acta Physiol Scand 17:440-451

Tesch P 1980 Muscle fatigue in man with special reference to lactate accumulation during short term intense exercise. Acta Physiol Scand suppl 480

Thorstensson A 1976 Muscle strength, fibre types and enzyme activities in man. Acta Physiol Scand suppl 443:1-45

DISCUSSION

Edwards: You raised the interesting question of differences in fatiguability between men and women. Another possibility, that takes us away from metabolism as the limiting factor, is that in women other factors are limiting exercise performance. One hypothesis that could explain your findings is that central fatigue is more evident in females than in males.

Karlsson: I haven't concluded anything about the background to these differences; they might very well depend on genetic factors as well as others such as social aspects, maturation, and environment. We haven't investigated any of these possibilities or combinations.

Bigland-Ritchie: How reliable are human muscle biopsies in providing information about the fibre composition of the muscle as a whole? In muscles of other species there is much evidence of an uneven distribution of fibre types (Burke 1981). The distribution of fibre types in women may be different, or more variable, so that sampling from one location could be misleading.

Karlsson: We have no evidence of a different general distribution pattern of muscle fibre types in women.

We have found somewhat different mean values in the vastus lateralis muscle between the two sexes (women 50 ± 8 [1 SD], and men, 55 ± 12.9% Type I muscle fibres, $P<0.05$) (Komi & Karlsson 1979). Maximal isometric knee extension force (MVC) was also different (women 40 ± 10 kPa and men 57 ± 16 kPa) as well as the surface recorded electromyographic activity (EMG) which averaged in women, 281 mV s^{-1} and in men, 545 mV s^{-1} ($P<0.001$). Moreover, the time taken to reach 70% of MVC (referred to as force time) was almost twice as long in women as in men (748 and 376 ms, respectively: $P<0.001$). Taken together, these results indicate a different neuromotor control in the two sexes in addition to a different metabolic profile in terms of muscle fibre types, muscle enzyme activities, and so on. But again, we don't know which is the 'genotype' or the 'phenotype' in these respects. To put it another way, we don't know which comes first, the chicken or the egg!

Saltin: On the question of how homogeneous or heterogeneous skeletal muscle is in its fibre composition, there is a tendency for there to be somewhat more ST fibres deeper in the quadriceps muscle. No sex differences have been noted. We have looked closest at the biceps muscle; the ST and FT fibres in this muscle are evenly distributed throughout the muscle, and we saw no differences between males and females in fibre type distribution (Nygaard & Sanchez, unpublished).

Karlsson: Your results are interesting in relation to Jack Wilmore's observation on the throwing ability of women and men (1979). According to this study, when testing the 'good arm' he found better performance in men, but equal performance capacity when testing the 'bad arm'.

Saltin: Both arms were studied and no difference was seen.

Roussos: To avoid the question of central fatigue differences between men and women, have stimulation studies been done to analyse the phenomenon?

Karlsson: We examined enzyme activities in muscles in males and females. A key enzyme is muscle phosphorylase. Its activity in females is 60% of that found in males (Komi & Karlsson 1979). We haven't looked at the activation of phosphorylase in women but we did this in men (Gollnick et al 1978). During muscular exercise there is, in males, no further activation beyond that already present at rest.

Edwards: Have you studied females as well as males, Dr Hultman?

Hultman: Yes. We did not find any difference between female and male subjects in our preliminary studies of glycogen phosphorylase and synthetase activities (D. Chasiotis, K. Sahlin & E. Hultman, unpublished observation). This agrees with the studies of Nuttall et al (1974) who found the same total activities of both phosphorylase and synthetase in male and female subjects.

They did observe an increase in synthetase I activity after glucose infusion in female subjects but not in male. The physiological significance of this sex difference is not obvious.

Edwards: If there is a difference in the absolute work rate, less lactate is likely to accumulate in the muscles of subjects with the lower work rate. Do you standardize for that, Dr Karlsson?

Karlsson: Yes. In some experiments we have tried to do so by using the exercise intensity or the oxygen uptake during a given exercise task—the bicycle ergometer, or the treadmill—versus a fixed lactate concentration corresponding to 4 mmol 1^{-1} blood as the measure of performance capacity. This is referred to as the lactate breaking point, lactate threshold, anaerobic threshold or, as we prefer to call it, *o*nset of *b*lood *l*actate *a*ccumulation (OBLA) (Sjödin et al 1980).

Newsholme: When we discuss fatigue and lactate we are presumably not saying that lactate itself is responsible for the fatigue, but the protons that are produced at the same time. There is no evidence that lactate in itself is harmful to muscle action. One must therefore take into account buffering capacity within the muscle. Does that vary from one individual to another, from one fibre type to another, and from males to females?

A second general point is that we must not over-interpret from a steady-state concentration. The concentration of lactate depends on its rates of production and utilization, and any change must take both these factors into account.

I am fascinated by the differences you find between males and females. I am not surprised that there are *quantitative* differences between the sexes, but I find it hard to believe in absolute differences in control processes. Could there be a different fate for the pyruvate that is produced, in females? Pyruvate can be converted to alanine, for example.

Karlsson: You raise the question of the driving factor in fatigue. There is a buffering capacity in the muscle tissue and to a certain extent lactate can be expected to be formed without any measurable change in pH. In our experiments we detect fatigue even at very low concentrations of lactate (true sub-maximal lactate concentrations) of 5–8 mmol kg^{-1} wet muscle. There is evidence from experiments on heart muscle that not only muscle tissue pH but also osmolality must be taken into account in discussing fatigue-induced changes in contractile properties.

Edwards: There are many products of muscular activity besides lactate. Dr Saltin has studied the accumulation of potassium in muscle which may significantly affect excitatory processes and could result in fatigue before maximal lactate concentrations are reached.

Stephens: Dr Karlsson, I was surprised that you plotted your results in terms of the percentage of fast twitch fibres (Figs. 4, 6, p 65, 68). We know that the fast twitch fibre population is a heterogeneous group mechanically (Garnett et al 1978). Some FT fibres are fatigue-resistant and some are highly fatiguable. The histochemical picture is not uniform, either. Are you missing something rather important here? Perhaps you should look more closely at the fast twitch population.

Karlsson: We have studied (Tesch 1980) the influence of FT sub-types on the fatigue pattern in males. Our results indicated, as expected, that individuals with a high percentage of FTb (Type IIB) fibres were more susceptible to fatigue than individuals with a higher percentage of FTa (Type IIA) fibres but with an equal ratio of FT to ST fibres.

I would certainly like to go further and see whether there is perhaps a sub-population missing in females but present in males, due to the effect of, say, testosterone. We have shown in castrated rats (Krotkiewski et al 1980) that testosterone is positively related to, among other things, muscle phosphorylase activity, and increases it. The mechanical and contractile properties of the muscles were not studied. The relevance of these observations to the present topic may seem obscure at this point, but at least it keeps the options open.

Edwards: Another major sex difference in enzyme handling is in the efflux from muscle of creatine kinase (Shumate et al 1979). Men lose more of this enzyme from their muscles during exercise than women. This might relate to what you are saying, because the efflux of creatine kinase is thought to be ATP-dependent. This difference has a practical implication in relation to the detection of women carrying the Duchenne muscular dystrophy gene (see Smith et al 1979).

Campbell: Can the striking difference between the sexes that you find in the speed of contraction be explained by selection of subjects?

Karlsson: The study (Komi & Karlsson 1979) was originally done to study the significance of monzygosity and dizygosity in relation to muscle fibre distribution and performance characteristics. The twins studied were from both Finland and Sweden. It was a randomized sample, in contrast to the exercise studies I referred to later in my paper. It is relevant to consider that we may be talking about differences in motor control and technique that are due to different interests among men and women in relation to sports. Again, these differences might have been developed at a very early stage in their maturation. G. Hedberg and E. Jansson have preliminary evidence (see Karlsson 1980) of differences between the sexes in the attitude to endurance exercise. The women studied do not in general have the same positive attitude as the men.

Edwards: Your twin studies suggested a dominant inheritance of slow twitch fibres. Was this seen only in males or did it apply equally to males and females?

Karlsson: It applied to both sexes equally.

Edwards: So there seems to be a separation between the genetic and the environmental influences. Perhaps you are now describing a change which is more environmentally influenced, or due to habitual activity, rather than to inherited factors?

Karlsson: This may be so. We see this difference in muscle fibre distribution (mean value as well as range) between females and males. Moreover, we have found this apparent difference in 'metabolic control' between the two main fibre types in females and males. Both possibilities are open to further studies, genetic and environmental.

REFERENCES

Burke RE 1981 Motor units: anatomy, physiology and functional organization. In: Brooks VB (ed) Motor systems. Williams & Wilkins, Baltimore (Handb Physiol sect 1 The nervous system vol 4), in press

Garnett RAF, O'Donovan MJ, Stephens JA, Taylor A 1978 Motor unit organization of human medial gastrocnemius. J Physiol (Lond) 287:33-43

Gollnick PD, Karlsson J, Piehl K, Saltin B 1978 Phosphorylase *a* in human skeletal muscle during exercise and electrical stimulation. J Appl Physiol 45:852-857

Karlsson J 1980 Localized muscular fatigue: role of muscle metabolism and substrate depletion. Exercise Sport Sci Rev 7:1-12

Komi PV, Karlsson J 1979 Physical performance, skeletal muscle enzyme activities, and fibre types in monozygous and dizygous twins of both sexes. Acta Physiol Scand suppl 462:1-28

Krotkiewski M, Kral J, Karlsson J 1980 Effects of castration and testosterone substitution on body composition and muscle metabolism in rats. Acta Physiol Scand 109:233-237

Nuttall FQ, Barbosa J, Gannon MC 1974 The glycogen synthase system in skeletal muscle of normal humans and patients with myotonic distrophy: effect of glucose and insulin administration. Metabolism 23, 6:651-568

Shumate JB, Brooke MH, Carroll JE, Davis JE 1979 Increased serum creatine kinase after exercise: a sex-linked phenomenon. Neurology 29:902-904

Sjödin B, Linnarsson D, Wallensten R, Schéle R, Karlsson J 1980 The physiological background of onset of blood lactate accumulation (OBLA). In: Komi PV (ed) Proc Int Symp Sports Biology, Vierumaki, Finland. University Park Press, Baltimore, in press

Smith I, Elton RA, Smith WHS 1979 Carrier detection in X-linked recessive (Duchenne) muscular dystrophy: serum creatine phosphokinase values in premenarchal, menstruating, post-menopausal and pregnant normal women. Clin Chim Acta 98:207-216

Tesch P 1980 Muscle fatigue in man with special reference to lactate accumulation during short term intense exercise. Acta Physiol Scand suppl 480

Wilmore JH 1977 Athletic training and physical fitness. Allyn & Bacon, Boston

Effect of metabolic changes on force generation in skeletal muscle during maximal exercise

LARS HERMANSEN

Institute of Muscle Physiology, Work Research Institutes, P.O. Box 8149, Dep. Oslo 1, Norway

Abstract During vigorous, strong contractions there is a rapid decline in the mechanical output or tension development in skeletal muscle. Several studies have indicated that this rapid decline in force development (often referred to as fatigue), is caused by metabolic changes in the muscles. During brief intense exercise there is a rapid breakdown of phosphocreatine and glycogen and a concomitant increase in the lactate and hydrogen ion concentration. The muscle lactate concentration is increased from about $1–2 \, mmol \, kg^{-1}$ wet weight at rest before exercise to approximately $25–30 \, mmol \, kg^{-1}$ wet weight immediately after intensive brief exercise to exhaustion. The muscle pH (i.e. the pH of muscle homogenates) falls from about 7.0 at rest to approximately 6.4 at exhaustion. The changes in the concentrations of ATP, ADP, and AMP are small. It is suggested that the changes in intracellular pH might affect the force generation of skeletal muscle by two different mechanisms: (1) The fall in intracellular pH reduces the activity of key enzymes in glycolysis, thus reducing the rate of ATP resynthesis, and (2) the increased hydrogen ion concentration has a direct effect on the contractile processes, thus reducing the rate of ATP utilization. It is suggested that the increased hydrogen ion concentration might be the common regulator for the maximal rate at which ATP is being utilized and the maximal rate at which it is being resynthesized.

The tension development or mechanical output of skeletal muscle declines after a sufficiently intense and prolonged period of exercise. This reduction in force generation of skeletal muscle, often referred to as muscle fatigue, has been a subject for experiments for a long time. Already in 1807 Berzelius had suggested that 'the amount of free lactic acid in skeletal muscle is proportional to the extent to which it has been previously exercised', according to Lehman (1850).

Today, more than 150 years later, we know a good deal more about the

1981 Human muscle fatigue: physiological mechanisms. Pitman Medical, London (Ciba Foundation symposium 82) p 75-88

metabolic processes taking place in muscle during heavy work. However, there is still no universal agreement about the fundamental mechanism of the reduction in tension development, or fatigue, experienced during maximal exercise.

The purpose of this paper is to present briefly some recent results on metabolic changes in human skeletal muscle during maximal exercise of short duration. In addition, possible mechanisms by which metabolic changes can affect tension development in muscle will be discussed.

Metabolic changes in human skeletal muscle during maximal exercise of short duration

During maximal exercise of short duration several rapid and pronounced metabolic changes take place both in the muscles and in the blood (Hermansen 1971, Karlsson 1971, Saltin & Karlsson 1971). Some of the metabolic changes taking place in human skeletal muscle are described below.

In resting human skeletal muscle the concentrations of adenosine triphosphate (ATP), adenosine diphosphate (ADP) and adenosine monophosphate (AMP) are about 4.65, 0.95 and 0.105 mmol kg^{-1} wet weight, respectively (Vaage et al 1978). After intermittent maximal exercise (three times 60 seconds of exercise to exhaustion with four-minute rest periods in between) the corresponding concentrations were 3.40, 1.00 and 0.103 mmol kg^{-1} wet weight, respectively (Vaage et al 1978). Earlier studies on human skeletal muscle have reported similar results (Hultman 1967, Karlsson 1971, Edwards et al 1980).

In this connection it should be mentioned that water is taken up by muscle during maximal exercise. The changes in muscle water content can be expressed as the wet weight: protein ratio, or as the wet weight: dry weight ratio (Hermansen & Vaage 1977). During intermittent maximal exercise the wet weight:protein ratio increased from 5.34 ± 0.06 at rest before exercise to 6.10 ± 0.05 immediately after exercise. This increase corresponded to a 14% increase in muscle water content. During the same type of exercise the wet weight:dry weight ratio increased from 4.36 ± 0.08 at rest before exercise to 4.85 ± 0.09 immediately after the intermittent exercise. This corresponded to an increase in the water content of muscle of 11%. These changes in the water content of muscle during maximal exercise should be included in the interpretation of changes in the concentration of ATP and other metabolites. They mean that part of the reduction in the concentration of ATP during maximal exercise is due to dilution. However, the fall in ATP is larger than can be accounted for by an increase in the water content. Since there are small changes in the concentrations of ADP and AMP, this means that part of the ATP is lost. The lost ATP is presumably converted mainly to inosine monophosphate. It should also be pointed out that the ADP concentration relevant to muscle contraction is that in free solution, in contact with the myofilaments. Since a large fraction (perhaps as much as 90%) of the ADP is bound, most of it to actin, measurement of changes in the total ADP concentration during exercise may be of limited importance.

Taken together, there seem to be only small changes in the concentrations of ATP, ADP and AMP, apart from changes in free ADP (Dawson et al 1978), during intense muscular work. Yet the tension development of the muscle is markedly reduced. Thus, it seems unlikely that the large reduction in force generation can be accounted for by the relatively small changes in the concentration of ATP.

The concentration of phosphocreatine is about $17.0\,\mathrm{mmol\,kg^{-1}}$ wet weight at rest before exercise. It falls to about $3.5\,\mathrm{mmol\,kg^{-1}}$ wet weight immediately after maximal exercise. In a recent study by Spande & Schottelius (1970) it was suggested that isometric force development was directly proportional to the phosphocreatine concentration in mouse skeletal muscle. They therefore concluded that depletion of phosphocreatine stores was the cause of muscular fatigue. However, Dawson et al (1978), who investigated muscular fatigue by phosphorus nuclear magnetic resonance, obtained results that did not support the Spande and Schottelius hypothesis. Dawson et al (1978) showed that there was no obligatory proportionality between isometric force development and phosphocreatine concentration. This renders the conclusion of Spande & Schottelius (1970) doubtful. Furthermore, there seems to be no obvious biochemical basis for postulating that depletion of phosphocreatine stores is directly responsible for the reduction in tension development or fatigue.

Thus, we can conclude that there seems to be no simple direct relationship between the concentrations of the immediate substrates (ATP and phospho-creatine) available for contraction, and the tension development during brief intense exercise.

During maximal exercise there is a rapid breakdown of glycogen and a concomitant production of lactate in the muscles. In subjects performing intermittent maximal exercise (Hermansen & Vaage 1977) the muscle glycogen content decreased from about $90\,\mathrm{mmol\,kg^{-1}}$ wet weight at rest before exercise to approximately $40\,\mathrm{mmol\,kg^{-1}}$ wet weight immediately after exercise to exhaustion. Although there is a large and rapid breakdown of glycogen during brief intense exercise, only about half of the muscle glycogen stores are utilized. Muscle fatigue during brief intense exercise occurs long before the muscle glycogen stores are depleted. Thus, lack of muscle glycogen cannot account for mechanical fatigue in brief intense exercise.

The concentration of lactate in resting skeletal muscle is about $1\text{--}2\,\mathrm{mmol}$ $\mathrm{kg^{-1}}$ wet weight (Hermansen & Vaage 1977, Karlsson 1971) and the pH is about 7.0 (Hermansen & Osnes 1972, Sahlin 1978). During maximal work the rapid breakdown of glycogen leads to a concomitant production of lactate and hydrogen ions.

After maximal exercise the concentration of lactate has increased to 25–30 mmol kg^{-1} wet weight (Hermansen & Vaage 1977, Karlsson 1971). Together with the increased lactate production there is an equal increase in the hydrogen ion concentration. The hydrogen ions produced are taken up to some extent by the buffer systems of muscle (and blood). The muscle bicarbonate concentration decreased from 10.2 mmol per litre of muscle water at rest before exercise to about 3 mmol per litre of muscle water after maximal exercise (Sahlin 1978). Although this and other buffer systems take up most of the hydrogen ions produced, there is still a pronounced fall in muscle pH as well as in blood pH (Hermansen & Osnes 1972) during brief intense work. Several studies (Furusawa & Kerridge 1927, Hermansen & Osnes 1972, Sahlin 1978) have shown that the pH of muscle falls from about 7.0 at rest before exercise to approximately 6.4–6.5 after vigorous exercise to exhaustion. Thus, a number of studies have shown that there are large changes in the lactate and hydrogen ion concentration during maximal exercise. In the following section, possible mechanisms by which changes in the hydrogen ion concentration may affect force development in skeletal muscle, and thus performance, will be discussed.

Effect of acidosis on rate of resynthesis of ATP

It is generally accepted that the contractile mechanism directly results in the hydrolysis of ATP. Moreover, it can function with no other source of energy.

In the results presented above I indicated that the total amount of ATP stored in the muscle cell is very small. Since there is only a small decrease in the level of ATP, even during maximal exercise, there must be a good balance between the rate at which ATP is hydrolysed and the rate at which it is being resynthesized. Thus, if the rate of ATP resynthesis is reduced during maximal exercise, force development will be directly affected. The question is, does this happen during maximal exercise in man?

During maximal exercise of short duration, only a fraction of the ATP utilized is resynthesized by oxidative phosphorylation and from the breakdown of phosphocreatine. Under these conditions (partly anaerobic) a large fraction of the ATP resynthesis is due to glycolysis. Skeletal muscle is known to have a very high glycolytic capacity: the rate of glycolysis may be increased 100 times or more (Newsholme & Start 1974). Thus, both the breakdown of glycogen and the production of lactate and hydrogen ions are very rapid. In order to avoid destruction of acid-labile cell components, the production of lactate and hydrogen ions must be regulated. However, this means that the glycolytic ATP resynthesis is also regulated. A reduction in the rate of glycolysis automatically leads to a reduced rate of ATP resynthesis.

The rate of glycolysis is affected by several factors. The activity of the two key enzymes (phosphorylase and phosphofructokinase) is known to play an important role in this regulation. *In vitro* studies (Danforth 1965, Ui 1966) of these two enzymes have shown that they are markedly affected by changes in pH. At a pH of about 6.4, the activity of both these enzymes is almost completely inhibited.

As reported above, the pH of muscle homogenates from samples obtained immediately after maximal exercise to exhaustion is also about 6.4 to 6.5. In this connection the studies of Hill (1955) should be mentioned. He observed that the formation of lactic acid in muscle in response to electrical stimulation ceased when the internal pH fell below 6.3.

Thus, one possible mechanism which may account for at least part of the reduction in force development is as follows. Maximal exercise of short duration leads to a rapid breakdown of glycogen and a concomitant rapid production of lactate and hydrogen ions. A fraction of the lactate and hydrogen ions produced during exercise diffuses out into the blood. As a consequence, the concentration of lactate in blood increases 20-fold (Hermansen 1971, Gollnick & Hermansen 1973) and the pH of blood may drop to 6.8. However, a large fraction of the lactate and hydrogen ions produced is 'stored' in the muscle cells. Although a large proportion of the hydrogen ions are buffered by the buffer systems of the cell, pH is falling. This intracellular acidosis will in turn reduce the activity of the glycolytic enzymes (i.e. phosphorylase and phosphofructokinase). A reduction of the enzyme activity will lead to a reduced rate of glycolysis and, thus, to a reduced rate of ATP resynthesis.

The explanation suggested above raises another question. If the development of tension in skeletal muscle is controlled by the rate of ATP resynthesis, why is there such a small reduction in the concentration of ATP? A possible answer to this question is given below (see hypothesis, p 81).

Effect of acidosis on the contractile processes

In a recent investigation Dawson et al (1978) studying isolated frog muscles showed that the decline in force generation was proportional to the rise in hydrogen ion and free ADP concentrations. In the following, changes in hydrogen ion concentration and their possible effect on the force-generating processes will be discussed.

The introduction of the 'skinned fibre' technique has made it possible to study force generation in single muscle fibres (that is, in parts of one fibre) under standardized intracellular ionic conditions. Small bundles of fibres were obtained from rabbit skeletal muscle by blunt dissection in mammalian Ringer's solution. Samples of human skeletal muscle were obtained using the needle biopsy technique as described by Bergström (1962). Single fibres were isolated in silicone oil and stripped of their sarcolemmas in an infused bubble of relaxing solution, as described by Donaldson & Kerrick (1975). The skinned fibres retain longitudinal integrity of the contractile proteins, as shown by their ability to generate tension. Contraction and relaxation were induced by changing the Ca^{2+} concentrations in the bathing solution. Despite the presence of mitochondria and sarcoplasmic reticulum, the skinned fibres are not dependent upon or influenced by normal excitation–contraction coupling mechanisms, intracellular buffer capacity, or metabolism, since their internal Ca^{2+} concentration, pH and $Mg\,ATP^{2-}$ concentration are determined by the composition of the bathing solution (Donaldson et al 1978). Thus, the force-generating apparatus is functionally isolated, and its behaviour can be studied under known experimental conditions. The changes in tension development under various conditions were measured using a small photo-diode force transducer.

The magnitude of steady-state isometric force at each Ca^{2+} concentration was determined at pH 7.0 and 6.5—that is, the approximate variation observed in human skeletal muscle during maximal exercise (see above). The maximum force generation was lower in all fibres at pH 6.5 than at 7.0. However, the decline in maximum force generation was larger in the adductor magnus muscle of rabbit than in fibres from the soleus muscle. Thus, the two main fibre types seem to react differently to a fall in pH. The 'anaerobic' fibres or Type II fibres seem to be more affected by a fall in pH from 7.0 to 6.5 than the 'aerobic' or Type I fibres from the same species.

There are several possible mechanisms by which acidosis might exert its negative effect on force generation. On the basis of current theories of how muscle contracts, either the sarcoplasmic level of Ca^{2+} or the response of the contractile proteins to Ca^{2+} might be changed. For instance, it has been shown that sarcoplasmic reticulum binds more Ca^{2+} as the pH is lowered (Nakamura & Schwartz 1970). It has also been shown that the amount of Ca^{2+} necessary to produce a given tension is greater at lower pH (Robertson & Kerrick 1976). There is also evidence which suggests that an increased concentration of hydrogen ions exerts a direct effect on the contractile process itself (Katz & Hecht 1969). Hydrogen ions might reduce the effect of Ca^{2+} on troponin, and thus affect force development. A direct pH dependence of Ca^{2+} binding to isolated troponin has not been clearly established. However, an acidotic depression of the Ca^{2+} sensitivity of the force-generating apparatus has been observed in glycerinated dog papillary muscle

(Schädler 1967) and in skinned skeletal muscle fibres of frog (Robertson & Kerrick 1976) and rabbit skeletal muscle (Donaldson et al 1978).

Thus, on the basis of the results presented above, the reduction in tension development observed during maximal exercise in human subjects may at least partly be accounted for by a direct effect of pH on the force-generating apparatus in the following way. Maximal exercise leads to a rapid and large production of lactate and hydrogen ions. A large proportion of the hydrogen ions produced are buffered by the intracellular and extracellular buffer systems. Yet, the intracellular pH falls from about 7.0 at rest before exercise to approximately 6.5 immediately after maximal exercise to exhaustion. This increase in the hydrogen ions leads to a reduced maximum force generation. The mechanism(s) for this negative effect is (are) not fully understood.

A hypothesis

As discussed above, the reduction in tension development may be accounted for by: (1) a reduction of the maximum rate of ATP resynthesis as a result of reduced activities of glycolytic enzymes due to increased intracellular hydrogen ion concentration, or (2) a direct effect of the increased hydrogen ion concentration on the force-generating apparatus. Since the concentration of ATP is only slightly reduced, and there is a large reduction in tension development, this might indicate that hydrogen ions have a dual effect, both on contraction (i.e. utilization of ATP) and on the resynthesis of ATP. It might well be that the increased hydrogen ion concentration is the common regulator for the maximal rate at which ATP is being utilized and for the maximal rate at which it is being resynthesized.

REFERENCES

Bergström J 1962 Muscle electrolytes in man: determined by neutron activation analysis on needle biopsy specimens: a study in normal subjects, kidney patients, and patients with chronic diarrhoea. Scand J Clin Lab Invest 14 suppl 68:1-110

Danforth WH 1965 Activation of glycolytic pathway in muscle. In: Chance B, Estabrook RW (eds) Control of energy metabolism. Academic Press, New York

Dawson MJ, Gadian DG, Wilkie DR 1978 Muscular fatigue investigated by phosphorus nuclear magnetic resonance. Nature (Lond) 274:861-866

Donaldson SK, Hermansen L, Bolles L 1978 Differential, direct effects of H^+ on Ca^{2+}-activated force of skinned fibers from the soleus, cardiac and adductor magnus muscles of rabbits. Pflügers Arch Eur J Physiol 376:55-65

Donaldson SK, Kerrick WG 1975 Characterization of the effects of Mg^{2+} on Ca^{2+}- and Sr^{2+}-activated tension generation of skinned skeletal muscle fibers. J Gen Physiol 66:427-444

Edwards RHT, Young A, Wiles M 1980 Needle biopsy of skeletal muscle in the diagnosis of myopathy and the clinical study of muscle function and repair. N Engl J Med 302:261-271

Furusawa K, Kerridge PMT 1927 The hydrogen ion concentration of the muscles of the cat. J Physiol (Lond) 63:33-41

Gollnick PD, Hermansen L 1973 Biochemical adaptations to exercise: anaerobic metabolism. Exercise Sport Sci Rev, vol 1

Hermansen L 1971 Lactate production during exercise. Adv Exp Med Biol 11:401-407

Hermansen L, Osnes J-B 1972 Blood and muscle pH after maximal exercise. J Appl Physiol 32:304-308

Hermansen L, Vaage O 1977 Lactate disappearance and glycogen synthesis in humans after maximal exercise. Am J Physiol 233:E422-E429

Hill AV 1955 Influence of external medium on internal pH of muscle. Proc R Soc Lond B Biol Sci 144:1-22

Hultman E 1967 Studies on muscle metabolism of glycogen and active phosphate in man with special reference to exercise and diet. Scand J Clin Lab Invest 19 suppl 94:1-63

Karlsson J 1971 Lactate and phosphagen concentrations in working muscle of man. Acta Physiol Scand suppl 358

Katz A, Hecht H 1969 The early 'pump' failure of the ischemic heart. Am J Med 47:497-502

Lehman CF 1850 Lehrbuch der physiologischen Chemische. (2nd edn, 1851, Cavendish Society, London)

Nakamura Y, Schwartz A 1970 Possible control of intracellular calcium metabolism by $[H^+]$: sarcoplasmic reticulum of skeletal and cardiac muscle. Biochem Biophys Res Commun 41:830-836

Newsholme EA, Start CM 1974 The regulation of metabolism. Wiley, London

Robertson S, Kerrick W 1976 The effect of pH on submaximal and maximal Ca^{2+}-activated tension in skinned frog skeletal fibers. Biophys J 16:73A (abstr)

Sahlin K 1978 Intracellular pH and energy metabolism in skeletal muscle of man. With special reference to exercise. Acta Physiol Scand suppl 455:1-56

Saltin B, Karlsson J 1971 Muscle ATP, CP, and lactate during exercise after physical conditioning. Adv Exp Med Biol 11:395-399

Schädler M 1967 Proportionale Aktivierung von ATP-ase-Aktivität und Kontraktionspannung durch Calciumionen in isolierten kontraktilen Strukturen verschiedenen Muskelarten. Pflügers Arch Gesamte Physiol 296:70-90

Spande JI, Schottelius BA 1970 Chemical basis of fatigue in isolated mouse soleus muscle. Am J Physiol 219:1490-1495

Ui M 1966 A role of phosphofructokinase in pH-dependent regulation of glycolysis. Biochim Biophys Acta 124:310-322

Vaage O, Newsholme E, Grønnerød O, Hermansen L 1978 Muscle metabolites during recovery after maximal exercise in man. Acta Physiol Scand 102:11A-12A (abstr)

DISCUSSION

Wilkie: We should not forget that it was shown by Claude Bernard in 1877 and more than fifty years later brilliantly exploited by Lundsgaard that if the formation of lactic acid is prevented, skeletal muscle rapidly fatigues and goes *alkaline*, so acidification certainly cannot be the whole story. (For a fascinating account of these and many related developments by other scientists, see Needham 1971, especially p 85 *et seq.* and p 367 *et seq.*)

Edwards: Dr Hermansen's hypothesis certainly emphasizes the balance between the demand for and supply of ATP. It is a key concept that production and utilization of ATP are controlled by similar means.

Karlsson: How much of the ATP present in the resting muscle is available for muscle contraction?

Wilkie: You are thinking of some of it being hidden away in compartments?

Karlsson: Not 'hidden away', but sequestered in some way.

Newsholme: The evidence from nuclear magnetic resonance (n.m.r.) studies is that significant amounts of ATP are not bound to proteins (Hoult et al 1974). Whether it is sequestered within vesicles or membranes is a different question. We measured the ATP concentration in muscle from many species across the animal kingdom, and we found that it is remarkably constant from the extremes of white muscle to the extremes of red muscle (Beis & Newsholme 1975). This suggests that ATP is not significantly compartmentalized.

Wilkie: I agree with that. In our experiments using a variant of the n.m.r. technique we (Brown et al 1980, 1981) can show that the terminal phosphorus of ATP and the phosphorus of phosphocreatine are freely exchanging in the resting muscle. Between 25 and 50% of the ATP exchanges every second, even at 4 °C. This argues against much of the ATP being metabolically or spatially isolated.

Karlsson: Some 10–15 years ago many electron microscopists (e.g. Hackenbrock 1966) discovered that mitochondria could have different forms, or 'energetic states'. This was supposed to be related to the translocation of ATP quanta from the inner part of the mitochondria to the cytosol along an ionic concentration gradient. This would imply a potential energy gradient from the mitochondria via the cytosol and out to the site of ATP utilization. The different 'contractile' stages would play a part in this 'pumping' process. As I understand this, the translocation of ATP quanta is a difficult process and the mitochondria are loaded with potential energy which can be expressed as a high local concentration of ATP.

Wilkie: In frog muscle the space occupied by mitochondria is very small—1 or 2% of the volume—so unless the concentration of ATP within the mitochondria is enormous, they wouldn't produce much ATP, however hard they are squeezed!

Newsholme: The evidence from mitochondria isolated rapidly from intact liver cells is that only a small proportion of the ATP is present within the mitochondria, so this theory does not seem feasible.

Edwards: If ATP is being generated in many parts of a cell at different times—for example, by cytosolic glycolytic enzyme pathways—there may be an effective answer to Jan Karlsson's question. Thus the ATP produced in peripheral parts of the cell may be preferentially used for membrane functions rather than being available for the cross-bridge interaction. Is there any information on the way in which ATP is bound in those circumstances, to enable us to assess the relative importance of contractile activity as against involvement of ATP in membrane functions?

Newsholme: The K_m values are very similar for all these processes and are far below the ATP concentration of the cell. In other words, these ATP-utilizing systems should function effectively until the ATP concentration is less than 10% of other normal values. After this the concentration of ATP may fall to very low levels very quickly.

Sjöholm: We have done some experiments related to this (Sahlin et al 1981). We poisoned rat skeletal muscle with iodoacetate. In that situation ATP is used up and falls below 40% of its original level. Then the muscle ends up in rigor. But ATP is actually used.

Wiles: The parallel human experimental situation is myophosphorylase or phosphofructokinase deficiency, where, when force starts to fail, the muscle may be partially in a state of contracture but the muscle ATP concentration is not reduced (Edwards & Wiles 1980).

Edwards: There is a sampling problem, of course. You may have a few fibres that are extremely depleted of ATP, and a majority which, for reasons to do with problems of excitation, have not been capable of being driven to that extent, and are acting to obscure by dilution the depletion of ATP in fibres in rigor.

Dawson: Our findings in frog muscle are in good agreement with those of Dr Hermansen in human muscle, that the decline in force seems to be related to the products of ATP hydrolysis, including H^+. How do you relate this to studies such as those of Rapoport and co-workers (Vergara et al 1977, Nassar-Gentina et al 1978) or of Lüttgau and his co-workers (Grabowski et al 1972), who feel that fatigue is directly related to a process in the activation of contraction?

Hermansen: In heart muscle, tension development decreases very rapidly without any effect on the action potential of the heart muscle cells. This indicates that—at least in some situations—there is a chemical factor in the cell affecting tension development.

Dawson: I believe that in the particular studies I mentioned, changes in the membrane or action potential were also ruled out as being *the* process of activation that is affected. These workers would put it further on, in some process between the action potential and cross-bridge cycling.

Edwards: Surely Lars Hermansen has given us *an* explanation for impaired excitation–contraction coupling. We are therefore bridging the gap between the chemical demands and electrical phenomena.

Campbell: If you lower the pH in other ways, for example by a high level of CO_2, do you see the same metabolic changes? Secondly, if you manipulated the extracellular conditions of the muscle, either in the human or in the isolated preparation, would you show that part of the effect is at the level of the cell membrane?

Hermansen: A.V. Hill (1955) showed that one could have large changes in the hydrogen ion concentration of the extracellular fluid with very little or no effect on tension development in skeletal muscle. As soon as there were changes of hydrogen ion concentration within the cell, there was a large effect. We have not altered pH by changing the CO_2 concentrations in our subjects.

Campbell: I was thinking that changing the CO_2 level is an easy way of varying intracellular pH.

Moxham: We have experimentally produced a particular type of muscle fatigue—referred to as low frequency fatigue—in the adductor pollicis of normal subjects. This type of fatigue has been mentioned by Professor Edwards (p 8) and will be considered in more detail later by Dr Jones (p 178), Dr Wiles (p 277) and myself (p 197). This fatigue is thought to be due to a problem of excitation–contraction coupling. We hyperventilated the subjects with a view to removing CO_2 and raising intracellular pH. We were able to show increased force at low stimulation frequencies, thereby reversing this type of fatigue (J. Moxham, C.M. Wiles, D. Newham & R.H.T. Edwards, unpublished work 1980).

Wilkie: We are in the same difficult situation as Dr Hermansen, of course; we have various factors correlated together but we don't know which one causes something else. Experiments must be devised so that one factor can be varied independently, in order to disentangle the situation. Intracellular pH can be varied controllably by varying CO_2 concentration. This is one of the many things that we plan to do. We know that if the CO_2 concentration is raised to 50%, using isolated muscle, contraction becomes small, relaxation becomes extremely slow, and the action potential becomes very prolonged. We haven't yet done the necessary biochemical tests using n.m.r. to answer the question.

Hermansen: This would be difficult to do in the human! You need a very high CO_2 concentration in blood to change the conditions in the muscle cell. This makes it difficult to do what Dr Campbell suggests in human subjects.

Campbell: I wasn't suggesting that you try to raise the level of CO_2 in the intact human. I was thinking of manipulation of the extracellular fluid in the sort of way you have been referring to.

Macklem: Didn't N.L. Jones et al (1979) show that metabolic alkalaemia improved exercise endurance times, whereas acidaemia decreased them?

Campbell: Yes, but was the sort of exercise stress used comparable?

Hermansen: I know that work. Others have shown no effect. It is difficult to achieve that much buffering capacity in the muscle cells. You need very high concentrations of bicarbonate in the blood to change the buffering capacity within muscle cells.

Hultman: You can easily decrease blood pH by CO_2 breathing but it is difficult to decrease intracellular pH by this means.

Edwards: The reverse may be possible, as John Moxham said, but I take your point about how difficult it is to raise P_{CO_2} chronically.

Saltin: Does that mean that there was less lactate? If cellular lactate levels were lower, that could be an argument for what Dr Hermansen is suggesting.

Roussos: We have confirmed that force is decreased by adding CO_2 to a bath containing an isolated guinea-pig diaphragm. When we release P_{CO_2} the force returns to normal. We have done this in soleus and diaphragm muscle.

Karlsson: We repeated the experiments of Jones et al (1979) and found large individual variations in the response. When we related changes in resting blood-buffering capacity and pH to the changes in lactate accumulation in exercise we found a positive relationship, indicating that the greater the increase in resting blood-buffering capacity after experimental treatment, the more lactate could be accumulated during short, maximal exercise (I. Jacobs et al, unpublished work).

Roussos: The total amount of lactate in blood that we could produce during strenuous exercise of the inspiratory muscles alone was 3 mmol/l. This level of lactate was accompanied by a normal or higher than normal P_{CO_2}.

Edwards: The important point is the relatively small mass of the respiratory muscle compared with the rest of the body muscle. We showed that submaximal exercise, especially with large increases in ventilation, produced a 1.0 mM increase in lactate, which one might expect from the small amount of active muscle (Edwards & Clode 1970).

Karlsson: Lars Hermansen discussed the accumulation of water in active muscle tissue. It is relevant to speculate about the increased osmolality. Have you tried to calculate that? And secondly, if under similar conditions you give urea or mannitol and dehydrate the contracting tissue the same effects can be achieved as with buffers or with increased calcium concentration, indicating that just by varying the water content one can normalize the situation.

Hermansen: During maximal exercise there is a large increase in lactate concentration in skeletal muscle, of about 25–30-fold. That will change the osmolality. Lactate is of course only one of the metabolites in muscle which changes during maximal exercise, but lactate and H^+ are probably among those showing the largest changes. That is why changes in H^+ concentration may have a special significance. I know no obvious physiological or biochemical mechanisms by which osmolality could affect tension development. That is why we have been looking to pH as a possible explanation of at least part of the large reduction in tension development.

Newsholme: As a general point, calculations based on the maximal activity of phosphorylase and the total amount of glycogen in human muscle indicate that the blood hydrogen ion concentration could increase to 1 M in 20 seconds if phosphorylase and glycolysis were fully active. Dr Hermansen made the point that protons inhibit phosphofructokinase, but that is not strictly true. The inhibition occurs, but it is a specific mechanism of *control*, since the inhibition depends on modifying the inhibitory effect of ATP. In other words, the effect of protons can be viewed as a physiological feedback mechanism that prevents excessive rates of glycolysis and the dangerous acidosis that could result if maximum glycolysis went on for too long. In a similar manner, if maximum myofibrillar ATPase was allowed to function, without regeneration of ATP, the total muscle ATP would be used in a few seconds and the cell would rapidly die. So the view that both processes (ATP utilization and regeneration) are regulated through the same or related mechanisms is physiologically sound. The biochemical nature of this mechanism is of course a fascinating question.

Edwards: Professor Hill, could you tell us about the effects of CO_2 on the mechanical properties of muscle? You studied the effects of very high concentrations of CO_2 on their twitch characteristics.

Hill: Yes. I showed (Hill 1968) that the twitch of the frog's sartorius muscle can be slowed by lowering the pH with CO_2, but a very high concentration is needed. A solution containing 30 mM-bicarbonate has to be equilibrated with about 80% CO_2 to produce a marked slowing of the twitch. The pH inside the muscle fibres is then about 6.0. The twitch of striated muscle is unaffected by pH until it falls to that sort of level.

There is one other point. If there is a direct connection between pH and force fatigue one should find a close parallelism between the restoration of pH and of force during recovery. This is not seen. Force returns to normal more rapidly than does pH.

Edwards: This emphasizes the important point that during the fatiguing process many things seem to go together but there is a divergence during the recovery processes, many of which have different time courses.

REFERENCES

Beis I, Newsholme EA 1975 The contents of adenine nucleotides, phosphagens and some glycolytic intermediates in resting muscles from vertebrates and invertebrates. Biochem J 152:23-32

Brown T, Chance EM, Dawson MJ, Gadian DG, Radda GK, Wilkie DR 1980 The activity of creatine kinase in frog skeletal muscle studied by saturation transfer nuclear magnetic resonance. J Physiol (Lond) 305:84P-85P

Brown T, Chance EM, Dawson MJ, Gadian DG, Radda GK, Wilkie DR 1981 The activity of creatine kinase in frog skeletal muscle studied by saturation transfer nuclear magnetic resonance. Biochem J 195: in press

Edwards RHT, Clode M 1970 Effect of hyperventilation on the lactacidaemia of muscular exercise. Clin Sci (Oxf) 38:269-276

Grabowski W, Lobsiger EA, Lüttgau HCh 1972 The effect of repetitive stimulation at low frequencies upon electrical and mechanical activity of single muscle fibres. Pflügers Arch Eur J Physiol 334:222-239

Hackenbrock CR 1966 Ultrastructural bases for metabolically linked mechanical activity in mitochondria. J Cell Biol 30:269-297

Hill AV 1955 Influence of the external medium on the internal pH of muscle. Proc R Soc Lond B Biol Sci 144:1-22

Hill DK 1968 Tension due to interaction between the sliding filaments in resting striated muscle. The effect of stimulation. J Physiol (Lond) 199:637-684

Hoult DI, Busby SJW, Gadian DG, Radda GK, Richards RE, Seeley PJ 1974 Observations of tissue metabolites using ^{31}P nuclear magnetic resonance. Nature (Lond) 252:285-287

Jones NL, Sutton JR, Taylor R, Toews CJ 1979 Effect of pH on cardiorespiratory and metabolic responses to exercise. J Appl Physiol Respir Environ Exercise Physiol 43:959-964

Nassar-Gentina V, Passonneau JV, Vergara JL, Rapoport SI 1978 Metabolic correlates of fatigue and recovery from fatigue in single frog muscle fibres. J Gen Physiol 72:593-606

Needham, Dorothy 1971 Machina carnis. Cambridge University Press, London

Sahlin K, Edström L, Sjöholm H, Hultman E 1981 Effects of lactic acid accumulation and ATP decrease on muscle tension and relaxation. Am J Physiol, in press

Vergara JL, Rapoport SI, Nassar-Gentina V 1977 Fatigue and posttetanic potentiation in single muscle fibres of the frog. Am J Physiol 232(3):C185-C190

The glucose/fatty acid cycle and physical exhaustion

E. A. NEWSHOLME

Department of Biochemistry, University of Oxford, South Parks Road, Oxford, OX1 3QU, UK

Abstract The energy required for sustained exercise is provided by the oxidation of two fuels, glucose and long-chain fatty acids, which are stored as liver and muscle glycogen and adipose tissue triglyceride. The latter provides the largest energy reserve in the body; there is sufficient energy for about five days of continuous marathon running. Glycogen reserves, in contrast, are very limited and, at most, could provide energy for 100 minutes. Evidence is presented of a metabolic limit in the rate of fatty acid utilization, so that sustained exercise at a high power output requires the utilization of both fat and carbohydrate simultaneously. There is a regulatory mechanism by which fatty acid oxidation reduces carbohydrate utilization in muscle—the glucose/fatty acid cycle. This plays an important part in ensuring that marathon runners can continue beyond the theoretical limit of 100 minutes. Triglyceride is mobilized from adipose tissue as long-chain fatty acids and the oxidation of these by muscle reduces the rate of glucose utilization. The availability of fatty acids for oxidation as early as possible in exercise will allow the use of *both* fuels (fatty acids and glucose) for a longer period of time. Since it appears that fatigue occurs when carbohydrate reserves are depleted, reduction in the rate of glucose utilization by the oxidation of fatty acids is obviously beneficial. The ability of ultra-distance runners to exceed these limits poses interesting metabolic questions relating to exhaustion.

Knowledge of the factors involved in the provision of energy for muscle and those that might be involved in exhaustion has accrued largely through the work of physiologists and classical biochemists. However, in the past 25 years considerable effort has been directed towards understanding the control of the rate of individual reactions in a pathway and hence the flux through the pathway as a whole (see Newsholme & Start 1973, Newsholme & Crabtree 1976, Newsholme 1978, 1980a). Application of this knowledge and the basic metabolic principles that arise from this work to the problem of the control of fuel supply for exercising muscle leads to some new insights into the question of physical exhaustion. These principles will be discussed in relation to the control of fuel

1981 Human muscle fatigue: physiological mechanisms. Pitman Medical, London (Ciba Foundation symposium 82) p 89-101

supply and exhaustion in long-distance running, the marathon in particular. Although more information is available for this run than for any other, detailed biochemical and metabolic information is sadly lacking, especially on elite marathon runners.

The major fuels for muscle during the marathon (a race covering 42.2 km) are glucose plus fatty acids obtained from the bloodstream and glycogen obtained from within the muscle. The important limitations in the use of these fuels will be discussed below. From this it will be shown that these fuels should not be indiscriminately used but that the rate of utilization of one fuel should be controlled, in relation to the energy demand by the muscle and to the rates of utilization of the other fuels.

Fuels for the runner

An indication of the significance and the limitation of the fuels will be provided in this section.

The use of blood-borne glucose

An elite marathon runner expends energy at a rate of about 84 kJ min^{-1} during the race (Costill & Fox 1969). Since 1 g of glucose produces 16 kJ of energy on complete oxidation, the runner would use about 5 g of glucose each minute. Since the total quantity of glucose in the extracellular fluid is only 20 g, glucose must be released into the bloodstream to prevent serious hypoglycaemia. This glucose is released from the liver. Experiments with both man and other animals demonstrate that liver glycogen is depleted during sustained exercise (Hultman 1978). Since the total hepatic store of glucose is only 100 g this would suffice for about 20 minutes. This represents a rate of glucose utilization of about 1.0 μmol min^{-1} g^{-1} muscle. Although the maximum activity of hexokinase has not been measured in muscles of elite marathon runners, it is about 1.0 μmol min^{-1} g^{-1} in muscle of fit normal subjects (Newsholme 1978, Newsholme et al 1980). This suggests that the marathon runner could support the energy demands of muscle by using blood glucose alone, but only for a limited period. Liver can also produce glucose from non-carbohydrate sources—that is, gluconeogenesis—and release this into the bloodstream for utilization by the muscle. Experiments on human volunteers have established that gluconeogenesis does contribute glucose to the exercising muscle (Felig & Wahren 1975). Unfortunately, these experiments were done on fit normal subjects, not athletes, and the workload was only mild (30% of maximum oxygen consumption). The rate of glucose utilization by the active muscles of these volunteers was approximately 0.2 g min^{-1}, or only 4% of the capacity. If the rate of gluconeogenesis is not higher in the elite runner, its quantitative importance is questionable.

The use of muscle glycogen

In sustained exercise muscle glycogen is depleted gradually. Volunteers were exercised on a bicycle ergometer at such an intensity that they became exhausted after about 100 minutes. Glycogen was assayed in biopsy samples of muscle taken during the exercise period: glycogen was used over the entire exercise period and exhaustion set in when the stores were depleted (Hermansen et al 1967). From the glycogen content of human muscle it can be calculated that, if the total muscle glycogen could be used, it would provide energy for about 70 minutes of marathon running (Newsholme 1980b).

The use of blood-borne fatty acids as a fuel

The largest fuel reserve in the body is triacylglycerol (fat) which, in theory, could ensure the fuel supply for about 119 hours of marathon running. Triacylglycerol is stored in adipose tissue which is distributed diffusely throughout the body; for example, under the skin, around the major organs and in the peritoneal cavity. It is released from adipose tissue as long-chain fatty acids which are transported in the bloodstream to the muscles, where they are taken up and oxidized to carbon dioxide and water. The blood concentration of fatty acids increases during sustained exercise by 3- to 6-fold and may reach a concentration of about 2 mM (Table 1). A concentration of

TABLE 1 Plasma concentrations of glucose, fatty acids and glycerol during sustained exercise in man (data taken from Felig & Wahren 1975)

Time of exercise (min)	Concentrations in plasma (mM)		
	Glucose	Fatty acid	Glycerol
0	4.5	0.66	0.04
40	4.6	0.78	0.19
180	3.5	1.57	0.39
240	3.1	1.83	0.48

2 mM reflects the maximal capacity of the high-affinity binding sites for fatty acids on albumin. Higher concentrations will increase the free concentration of fatty acids but this will lead to the formation of micelles, which are known to be dangerous. They damage cell membranes, increase the rate of aggregation of platelets, cause inhibition of enzymes and in hypoxic conditions increase the risk of cardiac arrhythmias (see Newsholme 1976, Spector & Fletcher 1978, Cowan & Vaughan Williams 1980). Furthermore, it is unlikely that the other lipid fuels—blood triacylglycerol, muscle triacylglycerol or ketone bodies—are quantitatively important as fuels for the marathon runner (see Newsholme & Start 1973, Felig & Wahren 1975).

Despite the fact that fatty acids are known to be used by muscle in prolonged exercise and that in theory they could support marathon running for five days, exhaustion in elite marathon runners occurs in less than three hours during a competitive run. This suggests that oxidation of fatty acids alone cannot provide sufficient energy for the muscles. Evidence in support of this view is presented below.

Limitations in the use of fatty acids as a fuel for muscle

Whereas the energy requirements of the elite marathon runner could be supported solely by blood glucose (at least for a short period), this is not the case for fatty acids. The evidence that fatty acid oxidation alone cannot support the maximum power output of the elite runner is as follows.

(i) If the carbohydrate stores of the body are depleted by feeding subjects a high fat diet before exercise, a given level of exercise produces exhaustion considerably more quickly than in subjects on a normal diet and especially in comparison to those on a high carbohydrate diet (Christensen & Hansen 1939).

(ii) If the carbohydrate store in the muscle is increased, a given level of exercise can be maintained for a longer period of time*.

(iii) If the fatty acid concentration in the blood is artificially elevated before exercise, a given level of exercise can be maintained for a longer period of time (Hickson et al 1977). This manipulation ensures that the plasma fatty acid concentration is elevated at the beginning of exercise rather than 30–40 minutes later (Table 1).

(iv) In ultra-distance runners, studied during a 24-hour run, the energy expenditure gradually declined from 87.5% of the maximum rate of oxygen consumption after one hour to 44.4% after 24 hours (Davies & Thompson 1979). It seems likely that, during this run, the availability of carbohydrate was progressively reduced so that fatty acid eventually became the only available fuel. This would suggest that fatty acid oxidation alone can provide about 50% of the maximum aerobic power output.

---

*The dietary regime is as follows. Six days before the competition muscle glycogen levels are decreased by an exhaustive run; the subject then eats a low carbohydrate diet for three days and once again runs to exhaustion to deplete the glycogen; for the final three days up to the competition the runner eats a high carbohydrate diet. This cause a greater than normal increase in the amount of glycogen deposited in the muscles (Bergström et al 1967). This dietary regime (known as 'glycogen stripping' or 'supercompensation') was probably first used by the British marathon runner Ron Hill in preparation for the 1969 European Championship in Athens: Hill won the marathon. Although there is no doubt that many marathon runners believe that it is important to raise the glycogen levels before the race, the need for the three-day period on the low carbohydrate diet has been questioned. Costill (1980) considers that a very high carbohydrate diet for a few days before the race, without the previous low carbohydrate period, is sufficient for 'supercompensation'.

These lines of evidence suggest that in the early stages of a run, only glucose and glycogen are oxidized by the muscle, but that after perhaps 20 minutes both glucose and fatty acids are oxidized simultaneously. Although both fuels are known to be used by the muscle, the question can be asked why, if glucose oxidation can provide all the energy required by the marathon runner and since glucose is available at normal or near normal levels in the blood, is it not used in preference to fatty acids? (Indeed, glucose is present at a higher concentration in the blood than are fatty acids.) This, of course, would soon produce severe hypoglycaemia, so that it is physiologically important that it does not happen. It is prevented by the operation of the glucose/fatty acid cycle (see Randle et al 1963, Newsholme 1976).

The glucose/fatty acid cycle in prolonged exercise

The availability of fatty acids in the bloodstream during prolonged exercise will favour their oxidation by muscle and so reduce the rate of utilization and oxidation of both glucose and muscle glycogen. The mechanism of this regulation by fatty acids is as follows. Fatty acid oxidation in muscle raises the intracellular concentrations of the important allosteric regulators of glycolysis and pyruvate oxidation, acetyl-CoA, citrate and glucose 6-phosphate. An increase in the acetyl-CoA/CoA concentration ratio will inhibit pyruvate dehydrogenase and markedly reduce carbohydrate oxidation; an increase in the concentration of citrate will inhibit 6-phosphofructokinase and reduce the rate of glycolysis. This latter effect will raise the concentration of glucose 6-phosphate, which inhibits both hexokinase and glycogen phosphorylase, resulting in a reduction in the rates of glucose utilization and glycogen degradation. If the rate of fatty acid oxidation were reduced, exercise would demand a greater rate of utilization of the limited carbohydrate stores and result in hypoglycaemia; this has been observed when fatty acid mobilization is reduced by nicotinic acid (Carlson et al 1963). Indeed, with the widespread clinical use of β-blockers in man, it has been noted that this treatment leads to the early onset of fatigue (Simpson 1977). Since β-blockers inhibit the release of fatty acids from adipose tissue, the lower blood concentration of fatty acids and hence their lower rate of oxidation could be an explanation of the fatigue.

It should be noted that, although fatty acid oxidation reduces the rate of glycolysis and pyruvate oxidation, this is not a fixed reduction and the rates of these processes will be higher than at rest. However, the rates of glycolysis and pyruvate oxidation will be less than they would be if there were no fatty acid oxidation. The mechanism of regulation of glycolysis is such that if the intensity of exercise is increased, and there is no compensatory change in the rate of fatty acid oxidation, the rate of glycolysis will increase (Newsholme 1977).

The metabolic basis of fatigue in endurance exercise

It would appear, from the considerations outlined above, that for the marathon run and indeed for all other track events that depend upon aerobic metabolism, the rates of oxidation of both the blood-borne fuels (fatty acids and glucose) cannot provide energy at a sufficient rate to meet the demands of the muscle. Hence muscle glycogen must be used to supplement the blood-borne fuels and fatigue will occur when it is depleted*.

These metabolic considerations lead to the view that the marathon runner should run at such a rate that his glycogen stores in the muscles are depleted just as the race ends. If the race is completed with glycogen remaining in the muscle, the athlete could have run faster: if all the muscle glycogen is used before the end of the race, the athlete would depend solely on fatty acid oxidation and the power output would fall by perhaps 50%. In the marathon runner this would be considered to be fatigue. Indeed, in ultra-distance running the power output is gradually reduced so that for perhaps the last 12 hours of the run the muscles do obtain almost all their energy from fatty acid oxidation. In this condition, the rate of fatty acid oxidation must be sufficient to reduce markedly the utilization of blood glucose by the muscles, to prevent hypoglycaemia. This suggests that the capacity for fatty acid oxidation must be very large in the muscles of the ultra-distance runners.

Triglyceride/fatty acid substrate cycle

Since the rate at which fatty acids are oxidized depends in part on their blood concentration, the control of the rate of mobilization is of considerable

---

* It has been usual for physiologists to propose that the supply of oxygen is the chief limitation to exercise performance. This concept cannot explain the beneficial effects of increasing the amount of muscle glycogen. The limitation in the utilization of blood-borne fuels is likely to be the ability of the muscle to extract the fuels (given that the concentrations of glucose and fatty acids in the muscle cannot be increased) which depends on their rate of diffusion from the bloodstream to the muscle. The rate of diffusion depends on the diffusion distance and the concentration difference. The former depends on the blood supply (capillary bed) and it is likely that this has already reached its limit of development in the well-trained athlete. The size of the concentration difference depends on a high concentration in the bloodstream and a low concentration in the muscle. The fact that the plasma fatty acid concentration is rarely found to increase above 2 mM, even in severe sustained exercise (or pathological conditions), suggests that this is an upper physiological limit. The intracellular concentration of free fatty acids in human muscle is not known, but it is likely to depend on the amount of the fatty-acid-binding protein (Gloster & Harris 1977) and the activities of the fatty acid utilization enzymes. Since these have probably been increased to maximal levels in the well-trained athlete, there is little improvement that can be made in this area of metabolism. Thus an increase in the glycogen content of the muscle is the only (obvious) major metabolic modification that can be, and has been, made by the marathon runner.

importance in providing a sufficient concentration to maintain a high rate of oxidation in muscle. Changes in the levels of the hormones insulin, glucagon and the catecholamines during exercise are important in increasing the rate of fatty acid mobilization.

In adipose tissue the process of lipolysis occurs simultaneously with that of esterification, so that triglyceride is broken down to fatty acids, which are re-activated and re-esterified to form triglyceride. Thus a substrate cycle between triglyceride and fatty acid is present in adipose tissue. One function of this substrate cycle could be to provide a sensitive control mechanism for fatty acid mobilization. Thus, changes in blood levels of lipolytic and/or anti-lipolytic hormones and other lipolytic regulators could ensure that the rate of mobilization of fatty acids is the same as the rate of their utilization by the muscle, so that exercise causes neither an excessive increase in fatty acid concentration, which could be dangerous, nor too small an increase, which would permit glucose to be utilized and would result in hypoglycaemia.

It is possible that endurance training increases the capacity of the triglyceride/fatty acid substrate cycle in adipose tissue. Consequently, an increase in the rate of cycling that would be produced by changes in hormone concentrations before the exercise could provide a very sensitive control mechanism by which the rate of fatty acid mobilization could be modified by small changes in the level of fatty acids (Newsholme 1977).

REFERENCES

Bergström J, Hermansen L, Hultman E, Saltin B 1967 Diet, muscle glycogen and physical performance. Acta Physiol Scand 71:140-150

Carlson LA, Havel RJ, Ekelund L, Holmgren A 1963 Effect of nicotinic acid on the turnover rate and oxidation of the free fatty acids of plasma in man during exercise. Metabolism 12:837-845

Christensen EH, Hansen O 1939 Arbeitsfähigkeit und Ernährung. Skand Arch Physiol 81:160-175

Costill DL 1980 Metabolic responses and adaptations to endurance running. 4th International Symposium on Exercise. Exercise and Hormone Regulations, in press

Costill DL, Fox EL 1969 Energetics of marathon running. Medicine and Science in Sports 1:81-86

Cowan JC, Vaughan Williams EM 1980 The effects of various fatty acids on action potential shortening during sequential periods of ischaemia and reperfusion. J Mol Cell Cardiol 12:347-369

Davies CTM, Thompson MW 1979 Aerobic performance of female marathon and male ultramarathon athletes. Eur J Appl Physiol 41:233-245

Felig P, Wahren J 1975 Fuel homeostasis in exercise. N Engl J Med 293:1078-1084

Gloster J, Harris P 1977 Fatty acid binding to cytoplasmic proteins of myocardium and red and white skeletal muscle in the rat. A possible new role for myoglobin. Biochem Biophys Res Commun 74:506-513

Hermansen L, Hultman E, Saltin B 1967 Muscle glycogen during prolonged severe exercise. Acta Physiol Scand 71:129-139

Hickson RC, Rennie MJ, Conlee RK, Winder WW, Holloszy JO 1977 Effects of increased plasma fatty acids on glycogen utilization and endurance. J Appl Physiol 43:829-833

Hultman E 1978 Regulation of carbohydrate metabolism in the liver during rest and exercise with special reference to diet. In: Landry F, Orban WAR (eds) 3rd international symposium on biochemistry of exercise. Symposia Specialists, Florida, vol 3: 99-126

Newsholme EA 1976 Carbohydrate metabolism in vivo: regulation of the blood glucose level. Clin Endocrinol Metab 5:543-578

Newsholme EA 1977 The regulation of extracellular and intracellular fuel supply during sustained exercise. Ann NY Acad Sci 301:81-91

Newsholme EA 1978 Substrate cycles: their metabolic, energetic and thermic consequences in man. Biochem Soc Symp 43:183-205

Newsholme EA 1980a A possible metabolic basis for the control of body weight. N Engl J Med 302:400-405

Newsholme EA 1980b The problem of fuel supply for the marathon runner. Medisport 2:155-157

Newsholme EA, Crabtree B 1976 Substrate cycles in metabolic regulation and heat generation. Biochem Soc Symp 41:61-110

Newsholme EA, Start C 1973 The regulation of metabolism. Wiley, London

Newsholme EA, Crabtree B, Zammit VA 1980 Use of enzyme activities as indices of maximum rates of fuel utilization. In: Trends in enzyme histochemistry and cytochemistry (Ciba Found Symp 73) p 245-258

Randle PJ, Garland PB, Hales CN, Newsholme EA 1963 The glucose fatty acid cycle. Its role in insulin sensitivity and the metabolic disturbance of diabetes mellitus. Lancet 1:785-789

Simpson WT 1977 Nature and incidence of unwanted effects with atenolol. Postgrad Med J 53 (suppl 3): 162-167

Spector AA, Fletcher JE 1978 Transport of fatty acid in the circulation. In: Dietschy JM et al (eds) Disturbances in lipid and lipoprotein metabolism. American Physiological Society, Bethesda, p 229-250

DISCUSSION

Edwards: Are you really giving us an up-to-date explanation of the old adage that 'fats burn in the flame of carbohydrates' and that the cycling mechanism is telling us how big the flame is?

Newsholme: The concept that 'fats burn in the flame of carbohydrates' implies that carbohydrates are needed to metabolize fatty acids. That is wrong in the sense that probably most athletes need to use both fats and carbohydrates as fuels and that the limitation in the oxidation of fat is not overcome by the presence of carbohydrate, except by complete oxidation of the latter. The intriguing question is what happens in ultra-distance runners who, after perhaps 12 hours, cannot use very much carbohydrate since it is not available. Felig & Wahren (1975) claim that 30% of the energy for prolonged exercise can come from glucose formed in the liver by gluconeo-genesis. However, this is during mild exercise (30% of maximum V_{O_2}), and we need to know the importance of gluconeogenesis when exercise is 70–80%

of maximum $\dot{V}_{O_2}$. Unfortunately, this has never been measured. I would expect that under those conditions gluconeogenesis would be relatively unimportant.

Edwards: One shouldn't forget the contribution of gluconeogenesis from protein metabolism, however.

Newsholme: As I say, we don't know the importance of glucose production from gluconeogenesis in severe prolonged exercise in elite distance runners. It is possible that amino acid oxidation provides significant energy, or alternatively that changes in the blood concentrations of amino acids are used as a metabolic signal for muscle metabolism. The branched-chain amino acids are almost exclusively used by muscle and changes in their concentration might be used as an important signal for the body.

Saltin: One problem with the gradual increase in importance of the turnover of free fatty acids during exercise is that when you measure the respiratory quotient (RQ) during the strenuous exercise with which we are concerned (marathon running), at 70, 80 or 90% of maximum $\dot{V}_{O_2}$, the RQ values after the first 10–15 min remain surprisingly constant. There is not a gradual fall. In the older literature the RQ did fall, but the relative work load was probably quite low. In all studies where the relative work load is known and was above 60–70% of maximum $\dot{V}_{O_2}$, it is surprising how constant the RQ remained.

Newsholme: In the ultra-distance running studied by Davies & Thompson (1979) the RQ fell gradually during the 24-hour run.

Hermansen: In our studies the RQ remained above 0.9 during the whole period of exercise ($1\frac{1}{2}$ hours) (Hermansen et al 1967).

Saltin: I thought someone might argue against using RQ values, but we have enough evidence that there is not much of a change over $1–1\frac{1}{2}$ hours of intense exercise, which tells us that although there may be differences among different fibres the *overall* utilization of fats and carbohydrates stays about the same.

Hultman: McGilvery (1975) calculated the maximum power output by muscle from the utilization of different energy sources. The highest power output could be produced by using ATP and phosphocreatine, next highest by the use of glycolysis and glucose oxidation; the lowest power output came from fatty acid oxidation. The values varied from $3.0\,\mu$mol $\sim$ P g^{-1} s^{-1} for phosphagen splitting, to 1.0 for glycolysis, to 0.24 when fatty acids were burned.

Newsholme: That is not relevant to the living metabolic system, which does not necessarily 'see' metabolism as seen by the biochemist. Living systems adapt, simply by having a greater capacity of a given pathway to accommodate any intrinsic differences in the energy supply from the fuels. The highest capacity for glycolysis is found in muscles that use only anaerobic glycolysis for energy production. The same rate of ATP formation can be achieved by *complete oxidation* of glucose, but the capacity of the pathway is less than 10% of that of the anaerobic cycle.

Hultman: These calculations were from exercise studies in human subjects. The uptake of fatty acids from blood was thus included in the calculation.

Newsholme: It depends what you take as your base. A given weight of glycogen in the living organism is much less effective in terms of ATP production than is fat, because glycogen is stored with water.

Edwards: You mentioned the specific dynamic action of food as a manifestation of the cycling rate. Is there a differential effect according to whether you are metabolizing protein or fat or carbohydrates? The old work seemed to show that protein had a much larger specific dynamic action than carbohydrate.

Newsholme: We have no information to prove or disprove that. We would certainly argue that, in terms of the specific dynamic action of food, some of the increased oxygen consumption may be accounted for by cycling. We believe that part of the oxygen consumed in, for example, the specific dynamic action and oxygen debt is due to cycling after exercise. We have predicted that cycling is increased quickly at the beginning of exercise, is reduced during steady-state exercise, but carries on slowly for a longer period when exercise has finished. An important question is for how long the oxygen debt goes on. Most people stop making measurements after an hour. Benedict & Cathcart (1913) reported an increased oxygen consumption of 13% above resting level seven hours after exercise. We would like to believe that most of this 13% was due to an enhancement of the rate of cycling. That takes us back to the importance of cycling in terms of weight control, which could be a benefit of exercise.

Merton: What about body temperature after exercise?

Newsholme: We have never measured that, but it can be remarkably high (41 °C) in some marathon runners.

Merton: Won't that distort the interpretation of the results showing increased oxygen uptake some hours after exercise?

Newsholme: No. First, the body temperature falls to normal within an hour; and second, increased oxygen consumption *must* indicate increased energy expenditure at whatever temperature. The important question is the cause of the increased energy expenditure.

Pugh: In athletes both core temperature and O_2 intake fall to pre-exercise levels within an hour after maximum exercise. In 19 measurements on different occasions in five middle-distance runners, O_2 intake from the 40th to the 50th minute after five minutes of maximum ergometer exercise was 2.01% SE $\pm$ 4.07% less than resting O_2 intake measured over 10 minutes before exercise (Pugh 1967).

Karlsson: If I understand you correctly, Dr Newsholme, there is some optimum value for the free fatty acid turnover in a given subject.

Newsholme: Not so much an optimum turnover, but an optimum concentration. The higher the turnover rate the better, but that depends on the concentration, and the upper safety limit appears to be about 2 mM.

Karlsson: Caffeine raises serum free fatty acid levels.

Newsholme: Yes. When you exercise, the fatty acid level falls in the first 10–20 minutes, because fatty acids are not mobilized immediately. The muscles use carbohydrate initially, but progressively make a greater use of fatty acids. So if you take caffeine beforehand, that will stimulate fatty acid mobilization from the start of the exercise, thus providing fatty acids *and* glucose in the blood from the beginning.

Karlsson: You measured adipose tissue triglycerides and free fatty acids bound to albumin in blood. What about circulating triglycerides and their significance in terms of energy turnover?

Newsholme: Our limited evidence is that the capacity of that system is about 10% of the total oxygen consumption of muscle. So serum triglycerides will not be major contributors as a fuel for exercise.

Saltin: In the ultra-long distance run that you discussed (Davies & Thompson 1979) the runners came down to a point where they were exercising at approximately 50% of their maximum $\dot{V}o_2$. That fits well with our results on glycogen depletion patterns. Above that level, when the slow twitch (ST) fibres are greatly depleted of glycogen, fast twitch (FT) fibres are also depleted. In Denmark we have a run of 100 km which takes subjects about 10–14 hours. We didn't see more than half the muscle glycogen gone after that run, and the remaining glycogen was mainly found in the FT fibres. Thus in this type of exercise fatty acid mobilization is crucial.

Campbell: What hormones are increased in exercise?

Newsholme: The hormones that are known to increase are the catecholamines (adrenaline and noradrenaline), glucagon, and possibly glucocorticoids and growth hormone. Insulin falls strikingly during sustained exercise. We would like to believe that the catecholamines play a direct role in fatty acid mobilization, but perhaps their biggest role is increasing the cycling rate, to make it sensitive to the other hormones, particularly to the fall in the concentration of insulin.

Campbell: How much of the catecholamine release is centrally mediated and how much is locally mediated?

Newsholme: I don't know. The evidence from denervation studies suggests that up to 50% can be locally produced noradrenaline. But the species variation in the sensitivity of adipose tissue to hormones is enormous, so what is true of the rat may not be true of the human subject. Much of the work has been done on the rat.

Edwards: Do you feel that any of this regulation takes place through carnitine palmitoyltransferase? In the myopathy field we are interested in the extent to which inherited deficiency of this enzyme (Di Mauro & Melis-Di Mauro 1973) might be in microcosm what you are seeing in extended exercise in normal subjects. These patients are in trouble during exercise if they don't have an adequate carbohydrate supply.

Newsholme: The first and essential step—what we call the flux-generating step—is the triglyceride lipase in the adipose tissue. Without mobilization of fatty acids, nothing can happen. But then there must be control within the muscle. If you give two muscles the same concentration of fatty acid, if one works harder it takes up more fatty acid. We would argue that that control is at the level of carnitine palmitoyltransferase. The mechanism, and whether it could be modified to produce a better performance, is an intriguing question.

Edwards: The activity of that enzyme is increased by training, as Holloszy's group have shown (Molé et al 1971). So it is an inducible enzyme, and it is a normal component of mitochondria, which themselves can be induced by activity.

Newsholme: I agree that one of the most important factors in training is the increase in mitochondrial enzymes, particularly of the TCA cycle and carnitine palmitoyltransferase.

Saltin: Is the limitation on the utilization of free fatty acids more inside the cell than in their transport from the capillary into the cell?

Newsholme: No. I would argue that once you have overcome the problem of mobilization, the major limitation is transport of fatty acids into the cell. But that doesn't rule out control within the cell. We don't know what controls the ability to take up fatty acids—whether it is capillary size, capillary number, or size of fibres. All these factors could be involved.

Roussos: Have any experiments been done in which this uptake has been impaired and performance reduced?

Newsholme: It has been done in animals by *increasing* the fatty acid level with heparin, and this increases the endurance (Hickson et al 1977). Other experiments have been done in which the fatty acid level was *reduced* and performance was also effectively reduced.

Pugh: Can you say anything about ketonaemia, or more specifically ketonuria, in relation to the fatty acid cycling?

Newsholme: Most of the evidence suggests that ketone bodies are not important in endurance exercise. In the Stockholm firemen studied by Felig & Wahren (1975) ketone bodies contributed less than 1% to the fuel requirement during exercise. In diabetic patients, kept without insulin for 12 hours, the ketone body level went up. Ketone bodies would undoubtedly contribute then, but only about 4% of the energy requirement.

Pugh: My reason for asking is their relation to fatigue in exercise of long duration. I have studied 350 competitors in a 45-mile walk (unpublished observations) and looked for ketone bodies in their urine. The second half of the field developed ketonuria after about three hours. The winners did not develop ketonuria at all. Johnson & Rennie (1974) made the same comparison with instructors and their classes in military long-distance exercises. They found the same: the classes became ketonuric but not the instructors.

Newsholme: That would suggest not that ketone bodies themselves were the cause of the fatigue, but that exhaustion would be produced when the glycogen stores are exhausted. Ketone bodies are indicative of the problem rather than the cause. They indicate a low carbohydrate reserve very effectively.

REFERENCES

Benedict FG, Cathcart EP 1913 Muscular work: a metabolite study with special reference to the efficiency of the human body as a machine. Carnegie Institute of Research publication no 187, Washington DC

Davies CTM, Thompson MW 1979 Aerobic performance of female marathon and male ultramarathon athletes. Eur J Appl Physiol 41:233-245

Di Mauro S, Melis-Di Mauro PM 1973 Muscle palmityl carnitine transferase deficiency and myoglobinuria. Science (Wash DC) 182:929-931

Felig P, Wahren J 1975 Fuel homeostasis in exercise. N Engl J Med 293:1078-1084

Hermansen L, Hultman E, Saltin B 1967 Muscle glycogen during prolonged severe exercise. Acta Physiol Scand 71:129-139

Hickson RC, Rennie MJ, Conlee RK, Winder WW, Holloszy JO 1977 Effects of increased plasma fatty acids on glycogen utilization and endurance. J Appl Physiol 43:829-833

Johnson RH, Rennie MJ 1974 Athletic training and metabolism. New Scientist 21 November, p 585-587

McGilvery RW 1975 The use of fuels for muscular work. In: Howald H, Portmans JR (eds) Metabolic adaptation to prolonged physical exercise. Brauheuser Verlag, Basel p 12-30

Molé PA, Oscai LB, Holloszy JO 1971 Adaptation of muscle to exercise. Increase in levels of palmityl CoA synthetase, carnitine palmityl transferase, and palmityl CoA dehydrogenase, and in the capacity to oxidize fatty acids. J Clin Invest 50:2323-2330

Pugh LGCE 1967 Athletes at altitude. J Physiol (Lond) 192:619-646

Shortage of chemical fuel as a cause of fatigue: studies by nuclear magnetic resonance and bicycle ergometry

DOUGLAS WILKIE

Department of Physiology, University College, London WC1E 6BT, UK

Abstract The technique of nuclear magnetic resonance (n.m.r.) is briefly described to illustrate its use for estimating metabolite levels *in vivo*.

Our studies of fatigue in anaerobic frog muscle at 4 °C are described in relation to (a) force development, (b) speed of relaxation and (c) the switching on and off of glycolysis. Both (a) and (b) are closely related, though in different ways, to the concentrations of key metabolites. In contrast, (c) is not related to metabolite levels as such but to the events of contraction and relaxation.

A special n.m.r. technique (saturation transfer) has been used to study the creatine kinase system *in vivo*. The results show that this system is highly active and is in equilibrium in resting muscle. The free [ADP] is consequently only a small fraction of that found by analysis of muscle extracts.

Studies of human power production as a function of duration of exercise also indicate that it is shortage of chemical fuel that brings short- and medium-term exercise (0.1–10 min) to a halt. It is proposed to extend n.m.r. methods to human subjects in the near future.

A working hypothesis to account for fatigue is suggested in which both the contractile system and the activating system play a part.

The technique of nuclear magnetic resonance (n.m.r.)

This technique depends ultimately on the fact that certain atomic nuclei—^{1}H, ^{2}H, ^{13}C and ^{31}P are the most interesting biologically—have spin, and thus a magnetic moment, in addition to the mass and positive charge that all nuclei possess. When subjected to a static magnetic field, B_0 tesla, such nuclei orientate themselves in relation to it. If an oscillating magnetic field is applied

1981 Human muscle fatigue: physiological mechanisms. Pitman Medical, London (Ciba Foundation symposium 82) p 102-119

at right-angles to the static one the nuclei tend to resonate at a particular frequency ν_L where

$$\nu_L = kB \tag{1}$$

and k is a constant for each type of nucleus; for example, for ^1H, $k = 42.6$ and for ^{31}P, $k = 17.2$ megahertz per tesla (MHz/T) respectively. Thus in a 5 T superconducting magnet, ^{31}P resonates at about 86 MHz. As a result of this resonance the nuclei continue to emit radiation for a few milliseconds after the brief exciting radiofrequency pulse is switched off: this radiation is collected and frequency-analysed to provide spectra similar to those shown in Fig. 1.

Note that the field experienced by the nuclei, B, is not exactly the same as B_0. This is because of the influence of neighbouring electrons and of nearby magnetic nuclei. The difference between B and B_0 (up to $3 \times 10^{-5} \times B_0$ for ^{31}P) in different chemical compounds is the basis of the *chemical shift* which enables us to identify and measure ATP, phosphocreatine, inorganic phosphate (P_i), sugar phosphates, etc., and to estimate the internal pH and Mg concentration in *completely intact* tissues. In addition to being non-invasive, n.m.r. has an advantage over conventional techniques in producing many items of interrelated information simultaneously. The first studies on contracting muscle were published in March 1976 (Dawson et al 1976; fuller details are given in Dawson et al 1977).

The biochemical changes that accompany fatigue

These experiments were done with Dr Joan Dawson of University College, London, and Dr David Gadian of the Department of Biochemistry, Oxford. The spectra were obtained on the 7.5 T spectrometer in that department.

The results of a fatiguing series of contractions in frog gastrocnemius muscle (anaerobic, 4 °C, tetanized for 5 s every 5 min) are shown in Fig. 1. The diminution in isometric force developed and the slowing of relaxation that are so characteristic of fatigue are shown on the right-hand side. The accompanying biochemical changes are shown by the spectra on the left: phosphocreatine breaks down progressively and P_i increases; ATP remains constant until fatigue is very advanced, when it diminishes by about 25%; the sugar phosphates of the glycolytic pathway increase somewhat, but not prominently in this particular pattern of stimulation; and the P_i peak shifts progressively to the left, showing that the internal pH is diminishing as a result of lactic acid production.

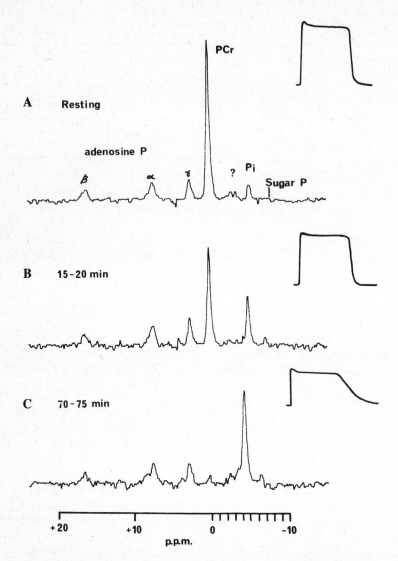

FIG. 1. Spectra and corresponding contractions obtained from two anaerobic gastrocnemius muscles stimulted at 5/300 s. The illustrations of force development have been traced by hand from the original records. PCr, phosphocreatine. P_i, inorganic phosphate.

A. Resting spectrum obtained after 2 h of anaerobiosis and just before the first stimulation. The peak labelled Sugar P at about −7.5 p.p.m. contains resonances from hexose and triose phosphates. AMP and IMP appear in the same general region. The ? refers to three compounds resonating at −2.7 to −3.6 p.p.m., whose functional significance is unknown. Neither the Sugar P nor the ? peaks showed consistent changes in size during the course of stimulation. The insert shows the first contraction of the series, obtained at time zero.

B. Spectrum obtained during the period 15–20 min after the commencement of stimulation. The insert shows the fifth contraction in the series, obtained at time $t = 20$ min.

C. Spectrum obtained during the period 70–75 min after the commencement of stimulation. The insert is the sixteenth and last contraction, obtained at $t = 75$ min.

Force development

Our studies of force development (Dawson et al 1978) have shown that the force is closely correlated with the concentrations of phosphocreatine, P_i, MgADP and H^+, regardless of the pattern of stimulation. This does not, of itself, distinguish between the two main—but not mutually exclusive—opinions about fatigue: (i) that it results from progressive impairment of the actomyosin ATPase system; and (ii) that it results from progressive failure of one or more of the links in the chain of activation from the central nervous system to the actomyosin system.

Certainly the products of ATPase activity—ADP, P_i and H^+—do increase markedly and might well slow down the essential biochemical process of contraction. This possibility could be tested *in vitro* on actomyosin systems but unfortunately we have not been able to persuade anyone with the necessary expertise to undertake the experiment. If failure of activation *does* play a part this must, in some as yet unknown way, be closely linked to the biochemical state of the muscle, a conclusion that was proposed some time ago by Lüttgau (1965).

Mechanical relaxation

Our studies on this topic (Dawson et al 1980a, b) have likewise shown that the rate constant for mechanical relaxation, $1/\tau$, where τ is the time constant, is a function of metabolite levels and is independent of the pattern of stimulation; however, the form of these relationships is quite different from that for force development.

Of particular interest is the accurately linear relationship between $1/\tau$ and the free-energy change per mole of ATP hydrolysed that can be calculated from our n.m.r. results. This quantity is important because (taken with negative sign and called affinity, A; Royal Society 1975) it measures the maximum work that can theoretically be obtained per mole ATP hydrolysed. If we are considering the actomyosin energy-transduction system, A is equal to the maximum amount of mechanical work that can possibly be obtained; when we are thinking of relaxation and the Ca^{2+} pumping that is firmly believed to bring it about, A equals the maximum work available to the Ca^{2+} pump for returning Ca^{2+} against a concentration gradient from the sarcoplasm to the sarcoplasmic reticulum. On present views (for details see especially Dawson et al 1980a) relaxation should just become impossible when $A = 39\,kJ\,mol^{-1}$. A slight extrapolation in Fig. 2 shows remarkable concordance between the *minimum* work required for calcium pumping if relaxation is to occur at all $(39\,kJ(mol\,ATP)^{-1})$ and the *maximum* work

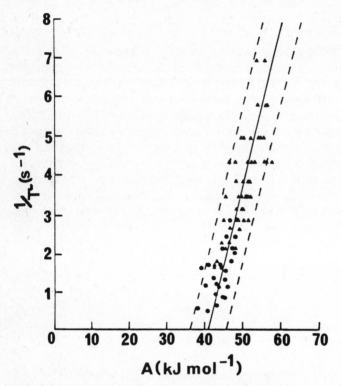

FIG. 2. $1/\tau$ as a function of free-energy change, or affinity, for ATP hydrolysis. Data represent the results of six experiments in which muscles were stimulated at 1s/20s, 1s/60s or 5s/300s. The initial and final parts of the experiments are represented by ▲ and ●, respectively. The continuous line is the regression of X upon Y obtained on all the points; the dashed lines are 95% confidence limits.

available from ATP hydrolysis $(41 \pm 5\,\text{kJ}(\text{mol ATP})^{-1}$, 95% confidence limits). Unfortunately our lack of knowledge about the details of the calcium-binding proteins, notably troponin C and parvalbumin, and their relation to force development during relaxation, makes it unfruitful to attempt to model the rest of the linear relationship shown in Fig. 2.

Recovery

Our early studies (Dawson et al 1977) confirmed that when oxygen is present the phosphocreatine broken down during contraction is completely rebuilt with a half-time of about 10 min.

Under the anaerobic conditions used in our experiments on fatigue the only source for the rebuilding of ATP and thus of phosphocreatine is the glycolytic formation of lactic acid. Though lactic acid cannot be seen directly in ^{31}P spectra, its production can be estimated from the observed changes in pH combined with knowledge of the buffers present. Incidentally, lactic acid can be readily measured in proton n.m.r. spectra, but despite a good deal of effort we have so far been unable for technical reasons to make accurate measurements on living muscles simultaneously at both frequencies.

Our results on anaerobic recovery are still being processed, though a preliminary account has been published (Dawson et al 1980b). They seem to show a remarkable uniformity: after widely different patterns of stimulation, roughly half the phosphocreatine that had been broken down is rebuilt within a minute or two, with corresponding production of lactic acid. After this the muscle remains completely quiescent for long periods, though further stimulation rapidly sets a similar process in motion once more (see Fig. 3). So

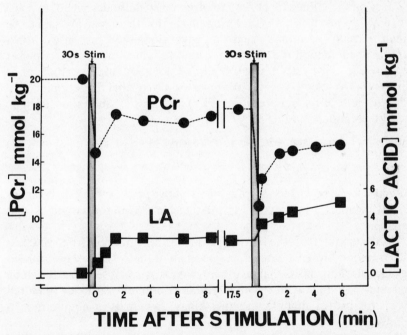

FIG. 3. The time course of changes in phosphocreatine (PCr) and lactic acid (LA) concentrations resulting from two 30 s tetani in anaerobic muscles. Note the breaks in the left ordinate and the abscissal scales.

far as the mechanism for the control of glycolysis is concerned, the interpretation appears to be clear. The switching on and off of the glycolytic pathway is very closely associated with the events of contraction, perhaps via the variations in internal calcium concentration. The substantial relative increases in the concentrations of ADP and AMP, which persist *between* contractions, do not activate the pathway. It remains something of a mystery that only half the phosphocreatine is rebuilt. We have sought in vain for a thermodynamic explanation and are now considering the evidence that there may be two populations of fibres, some of which recover completely and others not at all.

The creatine kinase equilibrium studied in intact muscle by saturation transfer n.m.r.

This work was done with three additional coworkers: Dr T. Brown of Bell Telephone Laboratories, Dr E. Chance of University College, London, and Dr G. Radda of the Department of Biochemistry, Oxford. A preliminary account has been published (Brown et al 1980).

The effect on the ^{31}P nuclei of the radiofrequency pulse described on p 103 takes several seconds to wear off. During this period the nuclei are said to be fully, then partially 'saturated', and they give diminished responses to test pulses. This phenomenon is a nuisance in that it limits how often test pulses can be applied; however, it can be turned to good use in attaching a short-lived label to selected peaks. We have used it to label the terminal (γ) P of ATP and the P of phosphocreatine (PCr) and thus to study the exchange catalysed by creatine kinase (EC 2.7.3.2):

$$MgADP^- + PCr^{2-} + H^+ \underset{\substack{\text{creatine} \\ \text{kinase}}}{\rightleftarrows} MgATP^{2-} + Cr \qquad (2)$$

This stoichiometry is correct at pH 7 and above: below pH 7 it becomes increasingly complex. For brevity, Mg and the charges will be omitted in what follows.

The results show that the exchange in resting muscle (frog, 4 °C) is extremely rapid; between a quarter and one-half of the ATP present is turned over per second. This shows that creatine kinase is active at rest (i.e. it is not a switched enzyme); and taken together with the observed constancy of [ATP] and [PCr] this proves that the creatine kinase system is at equilibrium in resting muscle.

This has many important consequences, one of which is that the concentration of *free* ADP which participates in reaction (2) is very much less than the concentration of *total* ADP that is obtained by chemical analysis after extracting the muscle with trichloroacetic or perchloric acid. There is conclusive independent evidence that in resting muscle almost all the intracellular ADP is firmly bound to actin and myosin (Carlson & Wilkie 1974, p 92).

Disregard for the distinction between total [ADP] and free [ADP] has led to confusion that need not have occurred. For example, Sahlin et al (1975) made an extremely thorough study of the creatine kinase system in man, using the needle biopsy technique to measure metabolite levels. They analysed their results as follows:

From (2),
$$\frac{[Cr][ATP]}{[PCr][ADP][H^+]} = K \tag{3}$$

from which

$$pH = -\log\frac{[Cr][ATP]}{[PCr][ADP]} + \log K \tag{4}$$

That is, when pH is plotted against the log [Cr][ATP]/[PCr][ADP], then the result should be a straight line of negative unit slope with an intercept on the pH axis equal to log K. Using the analytical *total* [ADP] the authors found that the slope was not -1 but -0.42, which is physically meaningless. Their estimate of $K = 2.4 \times 10^7$ M^{-1} was very much smaller than the values determined *in vitro* by Noda et al (1954) and by Lawson & Veech (1979), which were 2×10^9 and 1.66×10^9 M^{-1} respectively. A good deal of the paper by Sahlin et al (1975) is devoted to discussion of reasons for this strange result which seemed to show that in humans the creatine kinase system was not in simple equilibrium. The authors were aware that as much as 90% of the ADP present might be bound (see their p 178–179); had they carried this information to its logical conclusion, the whole situation would have become simple. This is shown in Fig. 4, which is replotted from Table 1 of Sahlin et al (1975), but with 3.06 mmol/(kg dry weight) subtracted from all the [ADP] estimations. The experimental points do now fall close to a line of unit slope whose intercept gives an estimate for K of 2×10^9 M^{-1}, in close agreement with the *in vitro* estimations. Thus in humans, as in frogs, there is every reason to think that the creatine kinase system is indeed in equilibrium and that in resting muscle [ADP] must be about 20–30 μM. Biochemical schemes that rely on higher concentrations of ADP for purposes of regulation should now be abandoned.

Bicycle ergometer studies

The analysis of studies of human exercise using the bicycle ergometer leads to the same conclusion as that derived from the n.m.r. experiments on isolated frog muscle: it is shortage of chemical fuel that limits work performance and finally brings it to a halt. In connection with attempts at man-powered flight,

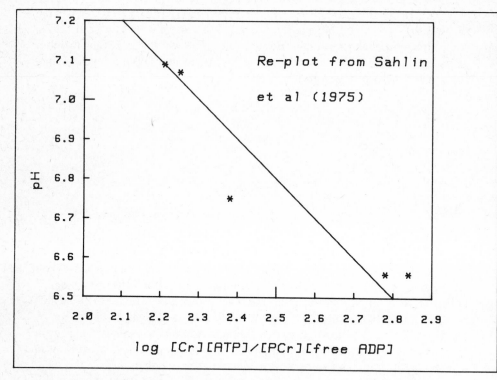

FIG. 4. Experimental results of Sahlin et al (1975, Table 1) replotted with 3.06 mmol/kg dry weight subtracted from all the ADP measurements to allow for the bound ADP. Three sets of experiments have been omitted on the independent criterion that in them [ADP] appeared to fall instead of to increase during exercise. This result could only arise from experimental error or from the intervention of another enzyme, such as adenylate kinase; there is no sign of the latter in the AMP measurements.

which has now been successfully accomplished, and at the request of the Royal Aeronautical Society, I undertook first a library search (Wilkie 1960), then (in collaboration with Dr Dawson; Dawson & Wilkie 1977) direct experimentation to discover the relation between mechanical power maintained (P, watts) and the total duration for which it was maintained (t, seconds). The subjects, four trained and experienced cyclists, were required to maintain a constant power level until they could continue no longer. Conditions were chosen so that exhaustion intervened in less than 10 min; beyond this duration the endpoint ceases to be really clear-cut.

Experimental results on one subject are shown by the stars in Fig. 5. The full line is drawn from the equation

$$P = E + A/t - E\tau[1\text{-}exp(-t/\tau)]/t \qquad (5)$$

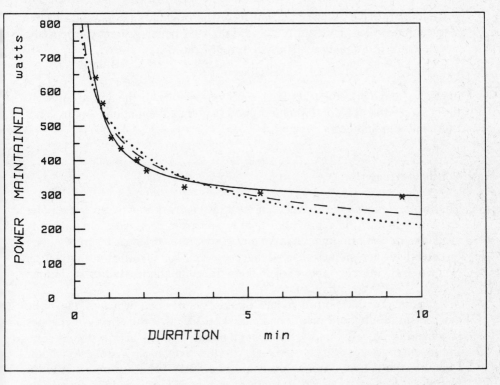

FIG. 5. Constant power maintained (watts, ordinate) in relation to total duration of exercise (minutes, abscissa) for subject G.D. The interrupted line has been fitted from equation (6) (Grosse-Lordemann & Müller 1937), the dotted one from equation (7), and the solid line from equation (5) (Dawson & Wilkie 1977).

Not only does this equation fit the experimental points very well, but also a physiological meaning can be assigned to the terms in it: E is the maximum *aerobic power* (273 W in this case); A is the total *work* available from anaerobic sources (16 kJ); the third term expresses the fact that the aerobic power does not rise instantly to its full value E at the onset of exercise but approaches it with a time-constant τ of about 10 seconds. This third term rapidly becomes negligibly small but it is nevertheless important in exercise of brief duration.

The interrupted line is drawn from the empirical equation of Grosse-Lordemann & Müller (1937):

$$t = 10^{(a \log P + b)} \qquad (6)$$

where a and b are constants without direct physiological significance. Clearly, the fit is not so good as that of the physiologically based equation (5).

It seems to have escaped the notice of the original authors, and of later commentators (e.g. Tornvall 1963), that the 1937 paper indirectly proposed a second empirical equation slightly different from equation (6):

$$t = [10^{(cP+d)}]/P \qquad (7)$$

in which c and d are constants. This equation is shown by the dotted curve. In this instance the fit is somewhat less good than that of equation (6); this is not true in all our subjects.

Concluding remarks

The two techniques that have been used in this work may seem to be poles apart but in fact they are not. N.m.r. spectrometers have now become available for work on intact human limbs, so in the very near future we hope to extend the techniques that we have evolved for investigating fatigue in isolated frog muscles and to use them in comparable studies of human exercise. [See Appendix, below.]

Returning to the central theme of this symposium, which is the basic mechanism of fatigue, I mentioned earlier (p 105) that the effects of changes in metabolite levels that we have demonstrated might act via the energy-transducing actomyosin system, or via the activation system, or both. Certainly the actomyosin system could not work if there were no ATP to hydrolyse. Observations on strongly motivated humans and on electrically stimulated frog muscles agree in showing that contraction diminishes and virtually stops before the ATP concentration has fallen very far, but when the back-up systems of phosphocreatine and glycolysis have reached their limits. This is just as well, for it means that neither athletes nor frog muscles can be stimulated to go into the rigor mortis that results from a substantial decline in [ATP]. I accordingly propose the speculative hypothesis, on little more than the notion that muscle is a well-designed machine, that though shortage of fuel acts primarily through the actomyosin mechanism, a second action on the activation system comes into play in extreme fatigue to prevent the muscle from destroying itself.

Appendix

During the week after the Ciba Foundation Symposium Dr Dawson and I had the opportunity to do the first experiments on human exercise using Topical ^{31}P n.m.r. The experiments were done at Oxford Research Systems (Ltd) using the 20 cm diameter spectrometer that had been built by them with the active participation of Dr D. Gadian, Dr G. Radda and Mr P. Styles of the Department of Biochemistry, Oxford University. During the experiments we worked closely with Mr I. Cresshole, Dr R. Gordon and Dr D. Shaw of Oxford Research Systems; Dr D. Gadian also participated in the experiments.

The movement studied was isometric hand-gripping between cylindrical pillars (diameter 25 mm) whose centres were 44 mm apart. Absolute forces are referred to the middle of the ulnar eminence.

The arrangement of the arm in relation to a surface coil and the shaped field of the superconducting magnet (200 mm bore, static field about 1.9 tesla) was such that ^{31}P spectra were obtained mostly from flexor carpi radialis and palmaris longus. Reasonable spectra showing phosphocreatine, ATP and P_i (if present) could be obtained over a five-minute interval using scans every 2 seconds.

Even at this early stage we feel confident that Topical n.m.r. has an important part to play in the study of human fatigue. For example, by a single experiment lasting about an hour we were able to show that in a series of 30 second maximal voluntary contractions repeated every five minutes, with the circulation occluded by a sphygmomanometer cuff, the force fell at the fifth contraction to less than one-tenth of its initial value of 273 N. At this point the phosphocreatine level had fallen only to 60% of its initial value. Both force development and phosphocreatine recovered rapidly when the cuff was released.

It is not clear at the moment whether this result conflicts with those already published on frog muscle at 4 °C (Dawson et al 1978), for subsequent surface recordings under similar conditions have shown that the EMG declines sharply in amplitude at the time when contraction virtually ceases. This suggests that in the human experiments the nerve, or some other part of the conducting pathway, fails. Obviously, a great deal remains to be done, and Topical n.m.r. seems to provide a good way of doing it.

REFERENCES

Brown T, Chance EM, Dawson MJ, Gadian DG, Radda GK, Wilkie DR 1980 The activity of Creatine kinase in frog skeletal muscle studied by saturation transfer nuclear magnetic resonance. J. Physiol (Lond) 305:84P-85P

Carlson FD, Wilkie DR 1974 Muscle physiology. Prentice-Hall, Englewood Cliffs, New Jersey

Dawson MJ, Wilkie DR 1977 Theoretical and practical considerations in harnessing man power. J Roy Aeronaut Soc Symposium on man-powered flight, March 1977

Dawson MJ, Gadian DG, Wilkie DR 1976 Living muscle studied by ^{31}P nuclear magnetic resonance. J Physiol (Lond) 258:82P-83P

Dawson MJ, Gadian DG, Wilkie DR 1977 Contraction and recovery of living muscles studied by
    ^{31}P nuclear magnetic resonance. J Physiol (Lond) 267:703-735
Dawson MJ, Gadian DG, Wilkie DR 1978 Muscular fatigue investigated by phosphorus nuclear
    magnetic resonance. Nature (Lond) 274:861-866
Dawson MJ, Gadian DG, Wilkie DR 1980a Mechanical relaxation rate and metabolism studied
    in fatiguing muscle by phosphorus nuclear magnetic resonance. J Physiol (Lond) 299:465-484
Dawson MJ, Gadian DG, Wilkie DR 1980b Studies of the biochemistry of contracting and
    relaxing muscle by the use of ^{31}P n.m.r. in conjunction with other techniques. Proc R Soc Lond
    B Biol Sci 289:445-455
Grosse-Lordemann H, Müller EA 1937 Der Einfluss der Leistung und der Arbeitsgeschwindig-
    keit auf das Arbeitsmaximum und den Wirkungsgrad beim Radfahren. Arbeitsphysiologie
    9:454-475
Lawson JWR, Veech RL 1979 Effects of pH and free Mg^{2+} on the Keq of the creatine kinase
    reaction and other phosphate hydrolyses and phosphate transfer reactions. J Biol Chem
    254:6528-6537
Lüttgau HC 1965 The effect of metabolic inhibitors on the fatigue of the action potential in single
    muscle fibres. J Physiol (Lond) 178:45-67
Noda L, Kuby SA, Lardy HA 1954 Adenosinetriphosphate-creatine transphosphorylase:IV.
    Equilibrium studies. J Biol Chem 210:83-95
Royal Society 1975 Quantities, units and symbols, 2nd edn. The Royal Society, London
Sahlin K, Harris RC, Hultman E 1975 Creatine kinase equilibrium and lactate content compared
    with muscle pH in tissue samples obtained after isometric exercise. Biochem J 152:173-180
Tornvall G 1963 Assessment of physical capabilities with special reference to the evaluation of
    maximal working capacity. An experimental study on civilian and military subject groups. Acta
    Physiol Scand 58:suppl 201:1-102
Wilkie DR 1960 Man as a source of mechanical power. Ergonomics 3:1-8

DISCUSSION

Macklem: What do you feel is the controlling mechanism by which the ATP concentration in the cell is not lowered below a critical level? Do you see it as a central reflex or something at the neuromuscular junction, or how do you envisage this?

Wilkie: Certainly not as a central reflex. Lüttgau (1965) showed that a suitable mechanism existed in isolated single frog muscle fibres. He showed that if you stimulate the fibres and obtained repeated twitches, the action potential becomes smaller and smaller, but if you put in metabolic inhibitors, so that the muscle no longer contracts, the action potential rises to its full size and stays there for several thousand action potentials. His main conclusion was that there is a connection between the contractile process and the size of the action potential. This is surprising, because the action potential doesn't depend on ATP hydrolysis for direct energy supply but merely for slow recharging of the ionic battery; so far as I know there are no big changes in ionic concentration accompanying fatigue. So protective control of contraction must be at the action potential or beyond. It doesn't have to be at higher levels.

Macklem: If you fatigue the respiratory muscles of a normal subject or an experimental animal there is a marked reduction in the duration of contraction. This is in a sense inappropriate, because you depart from the optimal tidal volume and frequency at which minimal work results. We have speculated that there may be a central reflex mechanism inhibiting the contraction of the inspiratory muscles, because if that did not occur they would eventually destroy themselves, as you pointed out. I can't think of any other mechanism for altering the duration of the inspiratory time than a central mechanism. You say that it's not *necessary*, but that doesn't exclude a central control.

Wilkie: You are right. A central mechanism *could* operate in this particular circumstance, in addition to the local one. But it isn't *necessary* to explain what is known about events in single fibres and isolated muscles.

Hultman: Thank you for recalculating our results using the free ADP! We (Sahlin et al 1975) found a good relation between

$$\log\left[\frac{(\text{creatine})(\text{ATP})}{(\text{ADP})(\text{phosphocreatine})}\right]$$

and pH in muscle samples obtained immediately after exercise when total ADP content was used in the calculation. If the same calculation was made using the content found after one minute of aerobic recovery the relation no longer fitted. A recalculation of these values using free ADP is difficult, as the total ADP content as a mean is lower (Table 2, Hultman et al, this volume, p 27) in the recovery phase than at rest. If the free ADP at rest is 10% of the measured value, what is the free ADP at increasing and at *decreasing* total content?

Wilkie: There is a very large increase, in relative terms, in the free ADP concentration as a result of contraction.

Newsholme: We studied creatine phosphokinase and also came to the conclusion that it is close to equilibrium, for quite different reasons. We also found this for arginine phosphokinase (Beis & Newsholme 1975). We studied 20–30 different species and on the same basis calculated the free ADP. We published the results on the kinase but not on ADP, because it was totally different in muscle from different animals. We did the same with 3'-phosphoglycerate kinase and also found tremendous variation from animal to animal. We also devised a system for measuring ADP directly at $-20\,^\circ$C using glycerol as a solvent, which indicated that about 50% of total ADP was bound.

There are problems if one says that only 10% of ADP is free at rest in muscle. Firstly, there is a problem of pyruvate kinase activity. The K_m of pyruvate kinase for ADP is about 0.1 mM, which is the concentration at which the enzyme exhibits half its maximum activity. At a maximum rate of glycolysis, pyruvate kinase must function at about half its maximum activity. This means that in such conditions the ADP concentration must approach 0.1 mM. Secondly, if you bind 90–95% of the total ADP, the ATP/ADP concentration ratio in muscle would be 60/1. Since there is a strict relationship between the cytoplasmic ATP/ADP ratio and the protonmotive force across the mitochondria, this would require an enormous protonmotive force *in vivo*. The highest ATP/ADP values found in isolated mitochondria are about 20:1 and, in the intact liver cell, about 5:1.

Wilkie: The advantage of our n.m.r. technique is that we do not have to use isolated mitochondria: instead we can see what happens in the intact living muscle. You mentioned earlier the very large control that must be exerted on glycolysis. The control exerted on oxidative phosphorylation between rest and activity is even greater, at least 3000:1. No one has ever achieved that with isolated mitochondria. If in living muscle we can show that creatine kinase is in equilibrium, and we know the equilibrium constant of the reaction, and if we know what the contents of phosphocreatine and creatine are, we can say what the free ADP level *must* be. There is no room for argument or speculation.

Newsholme: You need to use the *in vitro* system to measure K_{eq}, which will vary according to ionic strength, temperature and other factors, which you may not know *in vivo*. Also you need to know the distribution of creatine between extracellular fluid and other intracellular compartments, since creatine is not measured in the n.m.r. system.

Wilkie: We do this indirectly. We know from other experiments that in a fresh resting muscle the ratio of phosphocreatine to phosphocreatine plus free creatine is about 0.85 or 0.9. We also know that creatine doesn't diffuse out of muscles; in fact in our system it has nowhere to diffuse to. We can therefore say that the increase in creatine above its resting value is equal to the breakdown of phosphocreatine.

Edwards: Another possibility for studying this is to alter the substrates for the creatine kinase reaction. In some interesting studies, Dr Coy Fitch and colleagues in St Louis (Fitch et al 1974) fed β-guanidinopropionic acid to rats and replaced as much as 60% of their phosphocreatine with a phosphate receptor which seems to work like phosphocreatine. They have now looked at the contractile properties of muscle from such rats (Fitch et al 1978). The muscle seems to sustain force for longer before becoming fatigued. I don't know the equilibrium constant of the creatine kinase reaction with the new substrate, but might it give an indication of the relative importance of the reaction in fatigue?

Newsholme: I haven't seen that work. It is not just an academic point. The important point that Professor Wilkie is making is that if ADP is really 90% bound at rest, the ATP/ADP ratio is very high and muscle is then different from every other tissue. In liver it is 3:1, in adipose tissue, 2:1, and in kidney, 4:1. So skeletal muscle would be unique.

Jones: In the experiments by Fitch et al (1978) the soleus muscles of the treated animals maintained tension for longer than normal, although there was no difference in the function of the plantaris muscle. I understand from Dr J. Petrofsky that an intriguing aspect of these experiments is that replacement of phosphocreatine seems to result in a slowing of the muscle. Dr Petrofsky feels that phosphocreatine may be influencing the speed of the muscle in a way which is separate from its role supplying energy. It may be relevant to this that in rat, fast muscles have a higher phosphocreatine content than slow muscles (D. A. Jones & K. Gohil, unpublished observations).

Merton: It is still true, is it, that muscles go into rigor when the ATP level falls to a certain point?

Wilkie: Yes.

Merton: Is that the end? Can they not be taken out of rigor?

Wilkie: Not in the living muscle with an intact membrane. But if you have skinned the muscle fibres you can put on a relaxing solution that contains ATP and can reverse the rigor.

Jones: That is not true of all muscles. Isolated mouse and rat soleus muscles can be driven into a state of semi-rigor by repeated contractions so that they maintain about 20% of the full tetanic force (Fig. 5 of my paper, p 188, shows this beginning as the resting tension rises). If the muscle is allowed to recover in aerobic conditions the rigor tension decreases over three or four minutes.

Sjöholm: During stimulation, Professor Wilkie, is it possible to measure free ADP by comparing the α, β and γ n.m.r. peaks—their interrelations, as it were?

Wilkie: No, because even though the creatine kinase reaction is driven forwards during contraction by the large relative rise in ADP, the absolute rise is very small, too small to measure in our experiments. So we can't see ADP directly. You are right that it should show up as a diffference in the β and γ peaks.

Roussos: You consider the muscle basically as an engine that consumes chemical energy and produces work, and you explain fatigue on that model. How do you explain the fatigue due to neuromuscular transmission block or central fatigue? Is that something you exclude completely?

Wilkie: I don't exclude it; when people produce good evidence for fatigue further up the motor pathway, I certainly accept it! What I am saying is that you do see fatigue when you bypass the motor end-plate and work on an isolated single muscle fibre, as in Lüttgau's (1965) studies.

Roussos: Do you find the same in humans?

Wilkie: Not strictly. In humans, as was shown more directly in Eric Hultman's experiments, what brings the strongly motivated athlete to a halt is that he has run out of fuel; but the ATP level has not fallen much (Hultman 1967). Contraction fails when the back-up systems have run out of steam. Glycolysis is going as fast as it can, but not fast enough. Humans, like frogs, stop when the fuel runs out. In our experiments the contractions become very small, the chemical changes are also small, and the ATP level doesn't fall below 75% of its resting value.

Edwards: In the experimental design that you described you had short contractions with long recovery periods. Those contractions will give the best chance for energy to be depleted, rather than running into the problems with which many of us are involved in whole man, where there are impairments in either neuromuscular transmission or excitatory processes.

Merton: How did you stimulate the frog muscle in the n.m.r. studies?

Wilkie: With 50 Hz sine waves.

Merton: And when it started fatiguing, did you turn the stimulus up?

Wilkie: Of course!

Merton: And it did not recover at all?

Wilkie: No.

Jones: What happened to the action potentials?

Wilkie: We get 100 action potentials per second. There is no difference in fatigue or reproducibility between a 50 Hz stimulator or alternating condenser discharges (Csapo & Wilkie 1956).

Jones: But do you know what happened to the action potentials in the fatigued muscles?

Wilkie: I don't know. I do know that the rate at which fatigue comes on is the same whether you use 50 Hz sine waves or 10 cycles per second of alternating condensor discharges.

Edwards: I mustn't pre-empt the discussion of studies to do with the influence of frequency on excitation processes, but the possibility exists that failure of excitation, or excitation–contraction coupling, has occurred. In the absence of recordings of action potentials as the muscle becomes fatigued, we can't answer that point.

Wilkie: That is fair. I have recorded action potentials in unfatigued muscle and they do come where you would expect them to, at some point on each half cycle of the stimulating current (see Csapo & Wilkie 1956, p 496–499 for further details).

REFERENCES

Beis I, Newsholme EA 1975 The contents of adenine nucleotides, phosphagens and some glycolytic intermediates in resting muscles from vertebrates and invertebrates. Biochem J 152:23-32

Csapo A, Wilkie DR 1956 The dynamics of the effect of potassium on frog's muscle. J Physiol (Lond) 134:497-514

Fitch CD, Jellinek M, Mueller EJ 1974 Experimental depletion of creatine and phosphocreatine from skeletal muscle. J Biol Chem 249:1060-1063

Fitch CD, Chevli R, Petrofsky JS, Kopp SJ 1978 Sustained isometric contraction of skeletal muscle depleted of phosphocreatine. Life Sci 23:1285-1292

Hultman E 1967 Studies on muscle metabolism of glycogen and active phosphate in man with special reference to exercise and diet. Scand J Clin Lab Invest 19:suppl 94:1-63

Lüttgau HC 1965 The effect of metabolic inhibitors on the fatigue of the action potential in single muscle fibres. J Physiol (Lond) 178:45-67

Sahlin K, Harris RC, Hultman E 1975 Creatine kinase equilibrium and lactate content compared with muscle pH in tissue samples obtained after isometric exercise. Biochem J 152:173-180

Indirect and direct stimulation of fatigued human muscle

P. A. MERTON*[+], D. K. HILL[+] and H. B. MORTON*

*The National Hospital, Queen Square, London WC1N 3BG and [+]The Royal Postgraduate Medical School , London W12 0HS, UK

Abstract During voluntary fatigue in the human adductor pollicis muscle action potentials in the muscle evoked by stimulation of the opposite motor cortex did not diminish, so the whole motor pathway tested conducts normally. But twitches from direct stimulation of the muscle with massive pulses were greatly reduced, so, under the conditions used, it is the muscle fibres that fail.

> *Holmes, however, was always in training,*
> *for he had inexhaustible stores of*
> *nervous energy upon which to draw.*
>
> The Return of Sherlock Holmes

The natural and attractive idea that abundance of nervous energy can help us to overcome muscular fatigue nowadays looks very questionable. This paper describes the application to the study of human fatigue of two new techniques that emphasize the point. The first is electrical stimulation of the motor cortex. The method gives an immediate index of the responsiveness of the motor pathway as a whole, and promises to yield other information too.

The ready feasibility of cortical stimulation was discovered in January of 1980 (Merton & Morton 1980a, b). Two electroencephalogram electrodes or two pad electrodes are placed on the scalp, which is not prepared in any way. The stimuli are brief condenser discharges applied through an isolating transformer. Originally the stimuli were about 1500 volts and $10 \mu s$ time constant; we now find a time constant of $100 \mu s$ requiring voltages around 300 not obviously more uncomfortable. With the positive-going electrode over

1981 Human muscle fatigue: physiological mechanisms. Pitman Medical, London (Ciba Foundation symposium 82) p 120-129

the arm area of the motor cortex, twitches are obtained in the opposite forearm and hand. Action potentials can be led off over the contracting muscles. Their duration of about 20 ms shows that they are caused by a fairly synchronous motor volley. This, and the short and well-defined latency (about 16 ms in the forearm, 20 ms in the hand), suggest that the volley comes directly down the monosynaptic corticomotoneuronal pathway.

It is observed that the cortical threshold is much lowered if the muscle in question is contracted by voluntary effort. In experiments on voluntary fatigue this property has the useful effect of localizing the cortical stimulus to the muscles studied, provided the stimulus is not unnecessarily large. The muscle used here was the adductor pollicis. Force was exerted against a brass stirrup round the proximal phalanx and action potentials were picked up by surface electrodes on the palm, much as in Fig. 1 of Merton (1951, 1953), and using the original strain gauge. The force exerted was shown on a meter, to follow the progress of fatigue; it was not recorded.

As controls, records were first made of the response to single cortical stimuli, each given during a brief maximal voluntary contraction, with a minute's rest in between. Then the subject contracted the muscle as hard as he could for four minutes, the force falling to about a quarter of its initial value. With the subject still contracting, records of cortically excited responses were then taken at intervals of 10 s, the stimulus being exactly as before. Fig. 1 gives, for one experiment, the average of four control records and of four taken from the fatigued muscle. There is little to choose between them, from which it appears that the excitability of the motor cortex and the state of conduction in the motor pathway right through to the muscle fibre are not much affected by four minutes of maximal effort. This conclusion only applies to those muscle fibres and their central connections (most of them judged on the size of the muscle twitch) which are excited by the cortical stimulus.

The above experiment was done with the circulation to the hand free, so that blood flow, initially cut off by a maximal contraction, would have returned in the fatiguing muscle as the tension fell. When the experiment was done with an occluding cuff on the upper arm, cortically evoked action potentials became prolonged and smaller in peak height (to about half size) in extreme fatigue. This is evidence that the fibres excited are indeed those that are contracting to fatigue, confirming what was said earlier. Because of the change in the muscle action potential, however, the ischaemic experiment gives a less clear comparison with fresh muscle, for the purpose of evaluating the state of the motor pathway.

Another significant observation is that each cortically evoked action potential is immediately followed by a period of complete silence in the electromyogram (the start of which can be seen in Fig. 1) lasting several tens

Fresh

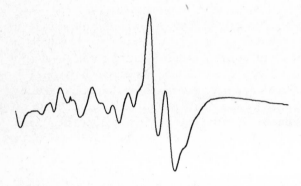

Fatigued

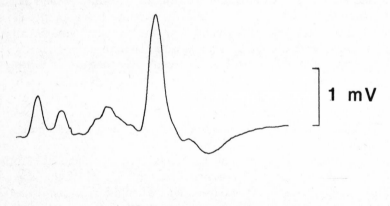

] 1 mV

100 msec

FIG. 1. Action potentials from the adductor pollicis with stimulation of the opposite motor cortex. The stimulus was delivered 20 ms after the start of the sweep and caused a response about 25 ms later. Each trace is the average of four. Four minutes of fatiguing contraction has little effect.

of milliseconds. The threshold stimulus for obtaining a silence is about as low as, and may be lower than, the threshold for excitation. A related observation was made by Penfield (1958). These facts could be read as evidence that the whole of the voluntary contraction in this muscle is excited or supported by the motor cortex, in the sense that there is no parallel pathway which can, on its own, keep the contraction going independently of the cortex. The technique is therefore probably telling us something which is relevant to the contraction under study. So far we have only learnt that the pathway remains wide open for single-shock stimulation in advanced fatigue. There are no hints about the more interesting question of how it is arranged that motoneuronal discharge slows during fatigue, as it must do if the muscle is to give of its best (Marsden et al 1976).

The second new technique is direct stimulation of the adductor pollicis, by-passing the motor nerves and the neuromuscular junction and giving unequivocal evidence about the state of fatigue in the muscle fibres themselves (Hill et al 1980). Gauze-covered sheet-lead electrodes soaked in saline are used, cut to cover the outline of the muscle and bent to fit the palm and the back of the hand. Large high-voltage pulses are passed by discharging a $0.25\,\mu F$ condenser, charged at up to $2400\,V$, through voltage-controlled silicon rectifiers (thyristors). The skin is not prepared in any way and does not suffer, not even reddening much. As the voltage is turned up the intramuscular nerves are stimulated first, as evidenced by a twitch starting about $4\,ms$ after the stimulus and preceded by only a small latency relaxation. At about $1000\,V$ the rising phase of the twitch jumps to an earlier take-off and a much larger latency relaxation develops, with a delay of only just under a millisecond. This is taken as evidence for direct stimulation of the muscle fibres. It is supposed that the latency relaxation becomes larger because it now occurs synchronously in the whole muscle. Above $1000\,V$ the size of the latency relaxation, the maximum rate of rise of tension at the start of the twitch, and the peak tension in the twitch all reach a plateau, presumably the result of stimulation of all the fibres in the muscle; what would normally be called maximal stimulation. At the highest voltages, however, the twitch agains increases in size (and duration), by a factor of 3 or 4, as seen on the plot in Fig. 2; but this large increase only affects that part of the twitch after the first $5\,ms$ or so; the latency relaxation and the early maximum rate of rise are not increased. The reason for this peculiar behaviour is not known. Possibly the latency relaxation and the initial fast rise of the twitch owe their invariance (above a certain voltage) to an association with the all-or-none spike potential of the muscle fibre. The exaggeration of the later parts of the twitch with massive pulses might be due to multiple firing or might depend on some process which is more susceptible to ill-treatment than the spike, such as the negative after-potential. At all events the largest stimuli (during which the

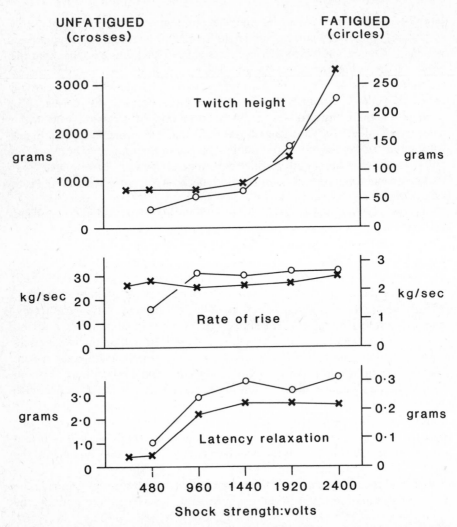

FIG. 2. Effect of four minutes of ischaemic fatigue on twitch height, early maximum rate of rise and latency relaxation in the adductor pollicis. The values for the fatigued muscle are scaled up 12 times, as indicated.

impedance of the preparation may fall to 35 Ω and the peak current rise to 40 A) do damage the muscle; even a few cause progressive elongation of successive twitches, an effect persisting for many minutes; and more than a few dozen may cause swelling which takes days to subside. No permanent harm is however apparent after a year's work.

These details are described because they are relevant to what happens in fatigue. After a maximal voluntary contraction for four minutes with the circulation arrested all elements of the mechanical response to direct stimulation are reduced by roughly the same amount. In the example shown in Fig. 2 the factor of reduction was close to 12. The latency relaxation, the peak rate of rise and the height of the twitch (both on the plateau and in the anomalous rise at very high voltage) all match the control values when plotted on 12 times the scale. The latency relaxation was difficult to measure on the fatigued twitches. On another occasion the factor of reduction was about 18. Twitches from the fresh and the fatigued muscle in this experiment are shown in Fig. 3. Apart from the obvious conclusion that the contractility of the muscle is reduced independently of neuromuscular block or any other feature of innervation, the significance of this result is uncertain. As the threshold to direct stimulation and the pattern of response of the various elements alter

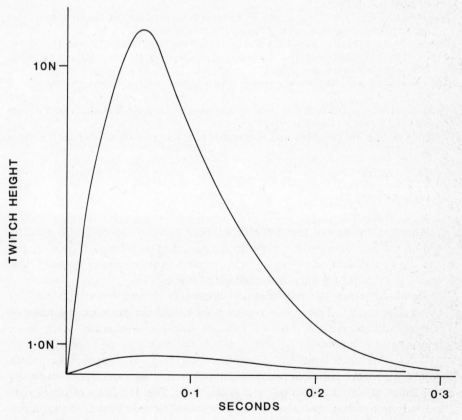

FIG. 3. Control and fatigued twitches with a shock strength of 1440 volts. From a different experiment to Fig. 2; the fatigued twitch was reduced 18 times.

little, it might be conjectured that the muscle action potential and the action potential–contraction coupling mechanism are not greatly affected by fatigue, but that they act on a weakened contractile system. This would also explain why the unphysiological giant twitches with very large shocks are also reduced in the same proportion in fatigue. But other explanations are almost equally plausible. On the experimental side, the possibility that the two distinct phases in the activation of a twitch, seen in the giant twitches, are also present in ordinary twitches is being pursued.

REFERENCES

Hill DK, McDonnell MJ, Merton PA 1980 Direct stimulation of the adductor pollicis in man. J Physiol (Lond) 300:2P-3P

Marsden CD, Meadows JC, Merton PA 1976 Fatigue in human muscle in relation to the number and frequency of motor impulses. J Physiol (Lond) 258:94P-95P

Merton PA 1951 The silent period in a muscle of the human hand. J Physiol (Lond) 114:183-198

Merton PA 1953 Speculations on the servo-control of movement. In: The spinal cord. Churchill, London (Ciba Found Symp) p 247-255

Merton PA, Morton HB 1980a Stimulation of the cerebral cortex in the intact human subject. Nature (Lond) 285:227

Merton PA, Morton HB 1980b Electrical stimulation of human motor and visual cortex through the scalp. J Physiol (Lond) 305:9P-10P

Penfield W 1958 The excitable cortex in conscious man. Liverpool University Press, Liverpool

DISCUSSION

Wilkie: You showed that the characteristics of contraction in the directly stimulated muscle were the same in fatigued and non-fatigued muscle, but you didn't mention relaxation. Is that slowed in the fatigued muscle? This seems to be almost a universal finding in fatigue.

Merton: The rate of relaxation is undoubtedly slowed, but not so much as one might expect. This is one conspicuous difference between the fatigued and unfatigued muscle; but we thought that the similarities were more interesting to emphasize.

Wiles: Although you have demonstrated that, at fatigue, the motor pathway from the cortex to the sarcolemmal membrane can respond normally to a single shock, this does not necessarily imply that the cause of the fatigue is solely peripheral. We also need to know whether, at fatigue, the motor pathway is capable of transmitting trains of impulses. It seems intuitively more likely that as the muscle's force-generating capacity declines, so the central nervous system modulates its outflow in parallel. There is considerable indirect evidence that firing rate is modulated from the earliest stages of a maximum voluntary contraction (see, for example, Dr Bigland-Ritchie's paper, p 130-148).

Merton: The most important problem, we think, is how the central pathway knows that the rate of discharge has to slow down, if the muscle is to be kept contracting. That point will be discussed later.

Edwards: It has to be established that you have 1:1 transmission from a shock to the cortex to the appearance of a single stimulus at the muscle itself. You might have triggered off multiple discharges at the spinal motor neuron.

Merton: That probably happens sometimes. The main action potential with moderate stimuli does not look as if there were multiple firing, but with a very big shock there is evidence of multiple firing.

Edwards: Also there is a big difference in the rise of tension in a singlet and doublet.

Merton: The twitches from stimulation at the periphery and stimulation at the cortex match remarkably well (Marsden et al 1981).

Lippold: J. W. T. Redfearn and I in the 1950s (not published) stimulated the motor cortex of the intact subject. Unlike Pat Merton, we found that the response was extremely variable and depended very much on the individual subject and thus on a lot of imponderables. Very often we could get a response in the form of a sub-maximal smooth contraction. We used frequencies of 30 Hz or thereabouts. The interesting point was that we found a latency of 10–20 seconds before obtaining any response from the muscle. The difference between this and your work must have been that we had to get the threshold of the pyramidal cells up and it takes that length of time to do it, whereas if you are making a voluntary contraction to begin with, you get a response with a latency of milliseconds only.

Stephens: If you stimulate the cortex soon after a maintained maximum effort begins, within the first 30 seconds of the contraction, what happens to the action potential in the muscle?

Merton: It gets bigger. We are not sure what the interpretation is. It may be trivial. Funny things happen to muscle action potentials in the first 20 seconds of a maximum contraction. The Piper rhythm gets bigger for a time before it diminishes. The appearance of the cortically evoked potential is rather similar. The explanation may be that if you have a volley which is slightly desynchronized, as a cortical volley is bound to be by the time it reaches the muscle, and, in the first few seconds of contraction, the muscle action potentials of the individual motor units widen, the peak-to-peak height of the compound action potential of the whole muscle could easily increase. So an increase does not necessarily reflect any change in the size of the volley.

Edwards: Did you stimulate the dominant or non-dominant hemisphere?

Merton: We have stimulated both sides of the cortex in the same experiment and there is no difference in latency to the start of the muscle action potential in the opposite arm that we can detect; it is probably less than a millisecond.

Stephens: You suggested that when you stimulate the cortex you get access to most of the neurons in the motor neuron pool. How much smaller is a maximal nerve-evoked twitch than a cortically stimulated, electrically evoked twitch?

Merton: The nerve-evoked twitch can be made within, say, 5% of the cortically evoked twitch, in rate of rise and peak height (Marsden et al 1981).

Stephens: That is an important point. It might be that different motor unit types behave differently during fatigue, as we shall hear from Dr Grimby. I wondered whether your cortical shocks are testing the more susceptible motor neurons in the periphery.

Edwards: Are you asking whether there might be a difference in sensitivity of the fast and slow motor unit pools, for example, at the spinal level?

Merton: It might be more complicated than I have made out. It is possible that we are reaching only half the fibres from the cortex, and they are firing twice, but it doesn't look like that, from the action potentials and the twitch shape.

Edwards: In the experiments in which you directly stimulated muscle there was a big difference in the size of the twitch from the fresh and fatigued muscle. The reduced twitch from the fatigued muscle could mean either that you had very little contraction from most or all the muscle fibres, or that some muscle fibres were still able to function but had stopped being recruited for central reasons. Can you say whether you are dealing with a proportional change in the functioning of all the fibres, or with different thresholds for stimulation in the fibre population of the adductor pollicis?

Merton: We cannot rule out the possibility that there are a few fresh fibres left, which have not fatigued; but I don't think it's very likely.

Jones: If the action potential remains the same size, the most likely explanation is that all the fibres are firing but not generating their full force.

Merton: We haven't faced the problem of recording action potentials immediately after passing very large current pulses straight through the muscle.

Wilkie: Is it feasible to tetanize the muscle directly? I can see a number of snags in doing so. I was wondering about the relationship of the twitch height to the tetanus height.

Merton: With the largest pulses the twitch tension is about half of P_0 (the tetanic tension).

Wilkie: The smaller the twitches are in relation to the tetanus, the more sensitive they are to all kinds of things that used to be called 'the active state duration' and such like.

Merton: It might just be possible to put in a tetanus.

Wilkie: You would have to be careful with unidirectional currents, presumably, because you are passing quite a lot of electric charge and moving a lot of ions.

Edwards: It is probably worth entering a note of caution here. I have been a subject of these experiments. Direct stimulation of the adductor pollicis is tolerable but it is a painful surprise! Before people go off and do these experiments, the size of the voltage has some relevance to the possible artifacts, due to the movement of calcium, presumably, by electrolysis. There is a method of bringing calcium into cells in isolated systems by punching holes in cell membranes with single large shocks (Baker & Knight 1978). Discrete small holes are produced in the membrane without damage to the rest of the cell. Through these holes, calcium can travel down its concentration gradient and bring about physiological responses—for example, stimulation of secretion.

Hill: The high voltage across the hand required to stimulate the adductor pollicis is nothing to be alarmed about. Almost all the volts are used up in forcing current across the skin. It is true that at the highest voltage used, 2400 V, the skin resistance breaks down so far that, as mentioned by Dr Merton, the current is great enough to have unphysiological consequences, but at 1000 V the quantity of electricity which passes through the muscle is no more than is generally used to stimulate an isolated muscle with transverse current. The voltage gradient across the muscle itself is too low to punch holes in the membranes.

REFERENCES

Baker PF, Knight DE 1978 Calcium-dependent exocytosis in bovine adrenal medullary cells with leaky plasma membranes. Nature (Lond) 276:620-622

Marsden CD, Merton PA, Morton HB 1981 Maximal twitches from stimulation of the motor cortex in man. J Physiol (Lond) (Proceedings November 1980), in press

EMG and fatigue of human voluntary and stimulated contractions

B. BIGLAND-RITCHIE

Quinnipiac College and John B. Pierce Foundation, New Haven, Connecticut 06519, USA

Abstract During a 60 s maximal voluntary isometric contraction (MVC) of the adductor pollicis muscle the loss of force is accompanied by a parallel decline in both the integrated surface electromyogram (EMG) and the single muscle fibre spike counts recorded intramuscularly. This decline is not due to neuromuscular block since the muscle mass action potential (M wave) evoked by single maximal shocks to the nerve is well maintained; nor does the size of the single fibre spike change. It must, therefore, reflect a decline in the firing pattern of the motor neuron pool.

The force of a sustained MVC continues to match that from maximal tetanic nerve stimulation; thus, all motor units remain active. Continuous nerve stimulation at the frequency required to match the voluntary force of unfatigued muscle leads to a progressive failure of the M wave, and a more rapid force loss than in an MVC. Both are largely restored by reducing the stimulus frequency. The decline in neural firing rate correlates well with the rate of muscle contractile slowing. It thus optimizes force by maintaining a relatively constant degree of tetanic fusion, while avoiding peripheral failure of electrical propagation.

A voluntary contraction depends on a chain of events starting with an adequate input to the motor cortex and terminating with the energy-dependent interaction of actin and myosin. Failure of any one of these events could result in the loss of force that characterizes fatigue (Fig. 1). Moreover, the particular site or combination of sites that fails first may well depend on the type and intensity of muscular activity causing the fatigue. These sites can be broadly divided into three categories: those concerned with delivering sufficient electrical activation from the central nervous system to the muscle; the various metabolic and enzymic processes providing energy to the contractile mechanism; and the excitation–contraction coupling processes that link these two. The experiments described here are concerned with the first of these categories. Traditionally, if force loss is accompanied by a parallel

1981 Human muscle fatigue: physiological mechanisms. Pitman Medical, London (Ciba Foundation symposium 82) p 130-156

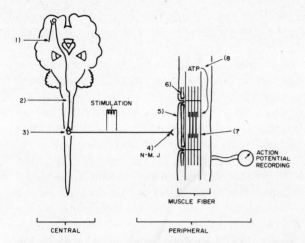

FIG. 1. Some possible sites of fatigue and the methods used to detect them.

1. Excitatory input to motor cortex
2. Excitatory drive to lower motor neuron
3. Motor neuron excitability
4. Neuromuscular transmission

5. Sarcolemma excitability
6. Excitation–contraction coupling
7. Contractile mechanism
8. Metabolic energy supply

decline in electrical activity, fatigue is attributed to failure of excitation; but if the electrical activity is undiminished the failure is attributed to the contractile system.

Many of the disagreements about the nature of fatigue arise from differences in methodology and interpretation when studies are made under different experimental conditions. Although fatigue, in most everyday activities, generally results from submaximal efforts executed intermittently and involving varying degrees of movement, we have confined our observations solely to fatigue during sustained maximal voluntary contractions for the following reasons. If fatigue is to be assessed in terms of force loss, this measurement must not be influenced by changes in muscle length or velocity of movement; nor can changing measurements of overall electrical activity be attributed to the changing behaviour of individual motor units if the number of active units varies. Moreover, both force and electrical activity measurements may be influenced by uncontrolled degrees of recovery if the activity is not continuous. Many of these complications are avoided in sustained maximal isometric contractions. How far our results apply also to other types of fatigue remains to be determined.

Most of our studies have been done on the adductor pollicis muscle because its contractions and electrical responses during voluntary effort can readily be compared with those resulting from stimulation of the ulnar nerve (Jones et al 1979, Bigland-Ritchie et al 1979, Bigland-Ritchie & Lippold 1979). Limited studies have also been done on other muscles including the quadriceps, which can be similarly activated by the femoral nerve or stimulated percutaneously (Bigland-Ritchie et al 1978).

Fig. 1 illustrates the principal sites at which fatigue may occur and the experimental approach we use to investigate them. To determine whether fatigue results from declining activation by the central nervous system, the rate of force loss during a maximal sustained isometric voluntary contraction (MVC) is compared with that from maximal nerve stimulation. If the force falls more quickly during voluntary activity and can be restored by nerve stimulation, fatigue is said to be 'central': if not, it must have resulted from failure at some site distal to the point of stimulation and is termed 'peripheral fatigue'. In addition, the effectiveness of electrical propagation across the neuromuscular junction and along the muscle surface membrane can be assessed during voluntary contractions by recording (a) the muscle surface mass action potential (M wave) evoked by single maximal shocks to the nerve, (b) the surface smooth, rectified (or integrated) electromyogram (EMG), and (c) the activity of a small sample of the total muscle, recorded from intramuscular fine wire electrodes. The latter allows any change in the overall muscle activity to be analysed in terms of changes in the number and size of individual fibre potentials, and sometimes the changing behaviour of the individual motor units. If force loss cannot be attributed to failure at any of these sites fatigue is deemed to result from events at, or beyond, the T (transverse) tubular system.

Fig. 2 shows typical records of force and EMG during a 60 s sustained maximum voluntary contraction of adductor pollicis. In addition, in many experiments M waves evoked by single maximal shocks to the ulnar nerve were recorded at 5 or 10 s intervals. The force starts to fall almost at once and continues roughly linearly so long as the muscle remains ischaemic—that is, until the intramuscular pressure no longer occludes the blood supply (Barcroft & Millen 1939, Merton 1954, Humphreys & Lind 1963). The rate of force loss is characteristic for each individual, usually amounting to 30–50% MVC per min for adductor pollicis (Bigland-Ritchie et al 1977) and somewhat more, 50–70%, for quadriceps (Bigland-Ritchie et al 1978). This rate is approximately inversely proportional to the subject's initial maximal voluntary force and may reflect individual variations in either the percentages of each muscle fibre type or their relative degree of hypertrophy.

The force loss is generally accompanied by a roughly parallel decline in the smooth, rectified (or integrated) EMG recorded either from the surface or intramuscularly. This decline in electrical activity could result either from a progressive blockage of electrical propagation across the neuromuscular junction and/or along the muscle fibre membrane, or from a reduction in the excitatory activation generated by the central nervous system.

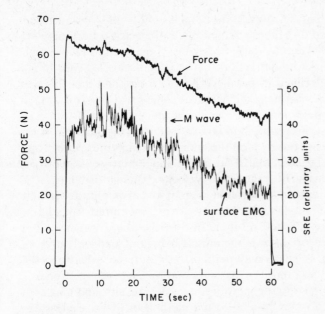

FIG. 2. Force and surface smooth, rectified EMG during a sustained maximum voluntary isometric contraction of the adductor pollicis muscle. Periodic recording of M waves evoked by maximal shocks to the nerve is also shown.

Neuromuscular block

When a nerve–muscle preparation is stimulated continuously, failure of propagation of electrical activity between nerve and muscle develops at a rate depending on the stimulus frequency (Thesleff et al 1959). This failure may occur: presynaptically at nerve terminal branches; postsynaptically, from a decrease in endplate excitability; or, less frequently, from a depletion of synaptic transmitter substance (Krnjević & Miledi 1958). Since transmission failure can be so readily demonstrated *in vitro*, and since similarly high rates of nerve stimulation are required to match the force of a voluntary contraction, it has often been assumed that neuromuscular block must be a principal cause of fatigue during voluntary contractions (Naess & Storm-Mathisen 1955).

However, there is an alternative point of view. The development of neuromuscular block is generally assessed by measuring the decline in the muscle compound mass action potential (M wave) evoked by single maximal shocks to the nerve. In 1954 Merton used this technique during voluntary fatigue of the adductor pollicis muscle; and he observed no decline in the amplitude of the M wave evoked periodically during the more than three minutes of maximal voluntary contraction despite almost total loss of force. He, therefore, concluded that there was no failure of electrical transmission.

In 1972 Stephens & Taylor challenged Merton's conclusion using a similar technique for contractions of the first dorsal interosseous muscle. They found that the surface EMG declined in parallel with the force, and that this was accompanied by a decline in the evoked M waves. These two observations led them to conclude that neuromuscular block was indeed the main cause of muscular fatigue at least during the first 60 s of sustained maximal effort.

In similar experiments of our own on the adductor pollicis muscle the total area of the M wave during a 60 s MVC, far from falling, increased somewhat, because of a slight slowing of conduction velocity (Bigland-Ritchie 1978, Fig. 1) and resultant increase in its duration without loss of amplitude. Similar results were also obtained when the M waves were measured from intra- muscular electrodes (unpublished results). In this case there was no doubt that the M waves were recorded from the muscle being fatigued and not from adjacent inactive muscles. There may perhaps be a difference between the fatigue mechanisms of these muscles, as suggested by Clamann & Broecker (1979). Alternatively, the discrepancy between our results and those of Stephens & Taylor (1972) may be due to the method of M wave measure- ment. We measure the total area of the M wave, above and below the isoelectric line, both by hand planimetry and by electronic integration.

However, even if the M wave area is well maintained, neuromuscular block might still play a role in fatigue of voluntary contractions. The M wave size is determined by the size of the individual muscle fibre potentials and the number of fibres that are activated.

$$M \text{ wave}_{(area)} \propto \left[\text{spike}_{(area)} \times \text{active fibre}_{(numbers)} \right]$$

During continuous high-frequency nerve stimulation (50–80 Hz) the area of the recorded potentials can double as a result of a profound slowing of conduction velocity before action potential failure becomes evident (see Fig. 4). If this also occurs during sustained voluntary contractions, up to 50% of the fibre could become blocked without a change in the size of the total M wave.

In more recent experiments we have simultaneously measured the M wave recorded from the muscle surface and the individual muscle fibre potentials recorded intramuscularly from fine wire electrodes. No detectable change was observed either in the waveform of spikes from single units identified and followed from the 15th to 60th second of contraction, nor in the summed areas of all recorded potentials divided by the spike counts (Fig. 8A). (Identifying single units during the first 10s of maximal contractions is seldom possible because of interference from surrounding units.) If neither spike size nor M wave area change, neuromuscular block as a cause for fatigue seems to be ruled out.

In the absence of neuromuscular block the declining muscle electrical activity during fatigue must result from a corresponding reduction in the discharge rate of the motor neuron pool.

Central versus peripheral fatigue

It is often believed that the central nervous system is not capable of fully activating a muscle during voluntary contraction; and also that in fatigue, a major source of force loss results from a progressive increase in this inability; that is, there is increasing 'central fatigue' (Mosso 1915, Reid 1928, Simonson 1971, Asmussen 1979). This belief is based more upon everyday experience and tales of superhuman strength and endurance in response to unusual psychological pressure, than upon firm experimental evidence.

When judging the functional capacity of the central nervous system during fatigue one must distinguish between what a subject thinks he can do (or is doing) and what his maximum capabilities really are. If a naive subject is simply told to pull as hard as he can for as long as he can, probably neither the initial force nor its sustained level will be maximal although he may well think he is doing his best. As his discomfort and sense of effort mount, there is also a progressive reduction of proprioceptive sensation so that, without visual feedback, he may be unaware that the force is actually falling. This is particularly evident when using a large muscle mass. As a result he does indeed start to 'let go' centrally. But to define this as 'central fatigue' and conclude that it is a necessary component of force loss in voluntary contractions requires proof that: (a) more force can be generated when the muscle's electrical activation is increased, as by direct stimulation; and (b) this *cannot* be done by increased effort on the part of the subject, if only briefly.

In our experience, maximum force can only be maintained reliably after considerable practice and with constant visual feedback. The force record is then so characteristic for each subject that any deviation becomes obvious. Despite these precautions some tendency to 'let off' still remains. In order to minimize this we constantly enhance the subject's motivation throughout each contraction by loud verbal encouragement and demands for periodic brief 'extra efforts'. These or similar precautions have not always been taken by those who conclude that 'central fatigue' is a necessary and fundamental part of the overall fatigue process in voluntary exercise.

Evidence suggesting that the central nervous system can indeed fully activate a muscle was provided in 1954 from two sources independently. Both showed that the maximum force of voluntary isometric contractions of the adductor pollicis muscle matched that of maximal tetanic stimulation of the ulnar nerve (Merton 1954, Bigland & Lippold 1954). Merton also concluded that this activation remained maximal when the contraction was maintained for three minutes, since nerve stimulation failed to restore the falling force. We have subsequently confirmed this observation many times. Interrupting sustained maximal contractions of adductor pollicis with brief periods of maximal tetanic nerve stimulation (Fig. 3A) has never resulted in an increase in force if the precautions for ensuring maximal force outlined above were observed. Central fatigue was also investigated for isometric contractions of the quadriceps muscle by comparing the voluntary force with either maximal femoral nerve stimulation (Fig. 3B) or percutaneous maximal stimulation of a constant fraction of the whole muscle (Fig. 3C and D). For three of the nine subjects studied the stimulated and voluntary forces fell in parallel—that is, there was no evidence of central fatigue; for three others the voluntary force always fell more quickly, and thus central fatigue seemed to be present (Fig. 3C); and for the remaining three the results were variable. But even those showing consistent evidence of central fatigue were able to regenerate force to match that from maximal stimulation during brief 'extra efforts' which they were unable to sustain (Fig. 3D). Thus central fatigue does not appear to be an insurmountable necessity.

The concept of central fatigue is supported by the work of Ikai et al (1967) who found a faster force loss during intermittent maximal voluntary isometric contractions of adductor pollicis than resulted from brief maximal nerve stimulation. Various methodological factors may explain their results. First, it is difficult to make sure each contraction is maximal when executed as fast as one per second, particularly when the muscle is slowed by fatigue. Secondly, they stimulated the ulnar nerve in the upper arm rather than at the wrist. The substantially greater force generated by nerve stimulation than in the voluntary contractions in the absence of fatigue suggests some cross-stimulation of additional muscle synergists innervated by the medial nerve, which were probably not used during the voluntary effort and thus remained unfatigued.

In general the available evidence, together with the more recent results presented by Merton in the previous chapter (Merton et al 1981, this volume), suggests that central fatigue is not a limiting factor for force generation during fatigue, at least of sustained isometric voluntary contractions. Furthermore, if the maximum force matches that of maximal tetanic nerve stimulation both before and after fatigue, then all motor units must be active and continue to contract with a fused tetanus throughout the contraction.

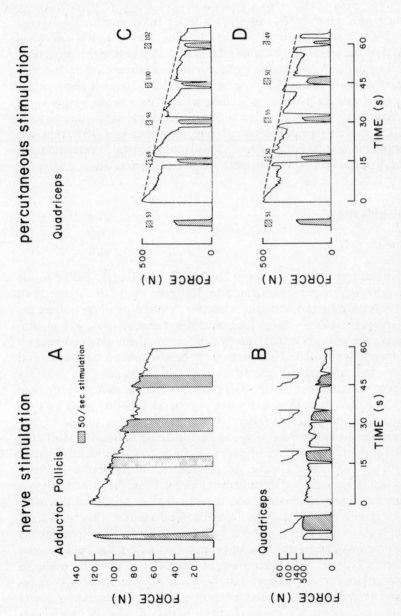

FIG. 3. Sustained maximal voluntary contractions of adductor pollicis and quadriceps muscles interrupted periodically by maximal tetanic stimulation of the nerve (A and B), or 53% (C) and 51% (D) of the total quadriceps muscle by percutaneous stimulation of intramuscular nerve endings. The contraction in C shows evidence of central fatigue. In D, the same subject was asked to make an 'extra effort' just before each stimulation period. The ratio between stimulated and voluntary force then remained unchanged as force declined.

Summary

The evidence presented so far suggests that in fatigue of sustained isometric voluntary contractions there is a reduction in the muscle's electrical activation; that this is not due to neuromuscular block; despite this reduction, there is little sign of central fatigue and the muscle continues to respond with a fully fused tetanus. These facts appear contradictory until changes in the muscle's contractile properties with fatigue and its altered response to different frequencies of activation are considered. To investigate this we have measured the muscle's response to different frequencies of activation and have attempted to imitate both the mechanical and electrical events of fatigue of voluntary contraction by various patterns of nerve stimulation.

Stimulated contractions

High frequency fatigue

The force–frequency characteristics of the human quadriceps and adductor pollicis muscles have been investigated by Edwards et al (1977). For both muscles a fully fused tetanus results when the femoral or ulnar nerves are maximally stimulated at 50–80 Hz (depending on temperature). Stimulation at 20 Hz generates only about 65% of the full force in the unfatigued muscle. When the muscle is stimulated continuously at this frequency, both force and M waves are well maintained despite some slowing of conduction velocity. With constant stimulation at 50 or 80 Hz the force falls far more rapidly than in an MVC, so that after 15 s more force is generated by low than by high frequency stimulation (Fig. 4). There is a marked slowing of conduction velocity, and the accompanying M waves first increase in duration and area before collapsing. If, at that stage, the stimulus frequency is suddenly reduced to 20 Hz, both force and M wave are restored to levels similar to those observed when 20 Hz was used throughout (Fig. 5). Thus, the rapid force loss during high frequency stimulation seems to be due to failure of electrical propagation rather than to defects in mechanical contractility or lack of energy supply, since the latter are both available once the action potential is restored. Force loss due to this type of propagation failure has been termed 'high frequency fatigue'. Although much of this failure may well be due to neuromuscular block in response to sustained high frequency nerve stimulation, reduced excitability and defective action potential propagation over the muscle fibre sarcolemma are also evidenced by the slowing of conduction velocity and the prolongation of the action potential waveform. Moreover, exactly similar responses to prolonged high and low frequency stimulation are

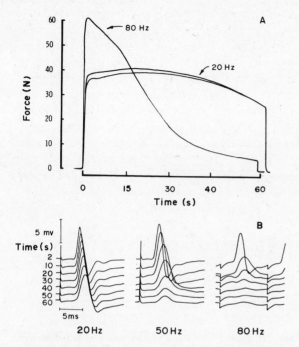

FIG. 4. Stimulation at various constant frequencies. (A) Force. (B) M waves evoked every 10 s.

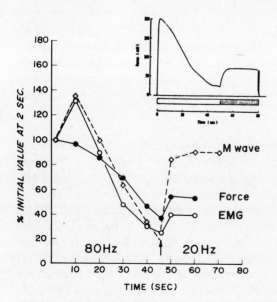

FIG. 5. Force, surface EMG and M wave areas recorded during maximum nerve stimulation at 80 Hz. At the arrow the stimulus frequency was reduced to 20 Hz. Inset: force record in response to similar stimulation of an isolated curarized mouse muscle.

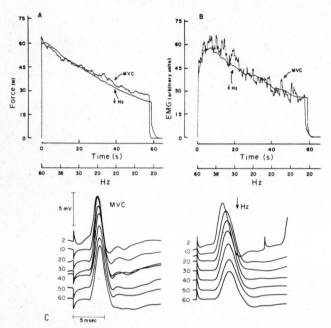

FIG. 6. A comparison of (A) force, (B) surface EMG and (C) M waves recorded during a sustained MVC and during maximal nerve stimulation in which the stimulus frequency was progressively reduced as shown.

obtained from directly stimulated isolated, fully curarized, anaerobic mouse muscle preparations from which neuromuscular transmission has been eliminated (Fig. 5, inset). This is discussed in detail in 'Fatigue due to changes beyond the neuromuscular junction' by D. A. Jones in this symposium.

Modelling voluntary contractions

Neither the rate of force loss nor the electrical activity of a sustained maximal voluntary contraction can be imitated by stimulating the nerve at any constant frequency. The best results are obtained when stimulating initially at 80 Hz, then rapidly reducing the stimulus frequency to about 20 Hz during the first 30 s (Fig. 6). This leads to an initial force, rate of force loss and rate of EMG decline that are similar to those observed in the MVC. (The absolute EMG levels differed because of the synchronized nature of nerve stimulation compared with asynchronous voluntary activity: Bigland-Ritchie 1978.) When the stimulus frequency is reduced in this manner the areas of the evoked M waves, like those in the voluntary contraction, also remained relatively constant in both shape and size.

Voluntary contractions

These results show that during nerve stimulation a progressive reduction of muscle activation frequency can optimize force generation in fatigued muscle by preventing neuromuscular block and other forms of failure of action potential propagation. The possibility that similar processes may also be involved during voluntary contractions has been investigated by periodically interposing brief periods of either high or low frequency maximal nerve stimulation during the course of a sustained MVC (Fig. 7). High frequency

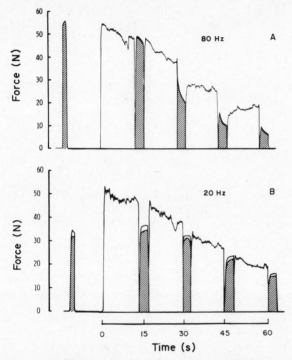

FIG. 7. A sustained MVC of adductor pollicis muscle interrupted periodically by maximal nerve stimulation (shaded areas). (A) At 80 Hz. (B) At 20 Hz.

stimulation matches the initial force but fails to do so after 30 s of voluntary fatigue; low frequency stimulation only generates 65% of the initial MVC force but matches it well after the first 30 s. Thus fatigue from voluntary contractions also seems to increase the susceptibility of the neuromuscular system to failure of action potential propagation when the activation frequency is high. The similarity between the voluntary force and that generated by low frequency stimulation during the latter part of contraction suggests that,

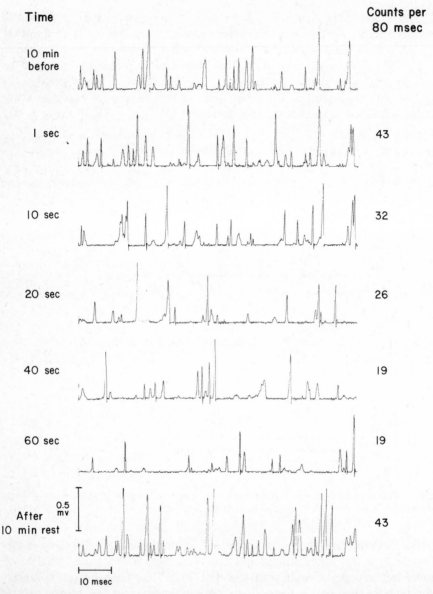

FIG. 8. (A) Intramuscular recordings of single fibre potentials taken at times shown during a fatiguing MVC.

as fatigue progresses, a reduction in motor neuron discharge rates may take place.

Direct measurement of changing muscle activation rate

In experiments designed to record directly changes in the firing pattern of the motor neuron pool during sustained maximal voluntary contractions, simultaneous recordings were made of force, integrated surface EMG and the frequency of single fibre potentials recorded from intramuscular fine wire electrodes. The total number of recorded potentials was counted both by hand and electronically. This included potentials from many different motor units.

Fig. 8A shows action potentials recorded periodically during a 1 min maximal voluntary contraction. As the contraction progressed, the mean frequency fell rapidly. Typical values were 60 spikes in 100 ms at the beginning, falling to 5–10 spikes after 60 s.

When mean firing frequency was plotted against time, a curve having two components was found (Fig. 8B). Initially, there was a rapid decrease in frequency, followed by a less rapid change. A sharp discontinuity of rate occurred at about 10–15 s. Both components were well fitted by straight lines. During the first 15 s there was an approximately 50% fall in rate, followed by a further 25% decline during the next 45 s.

The intramuscular electrodes were assumed to record the electrical activity of a small, but constant, sample of the muscle. The activity of this sample was

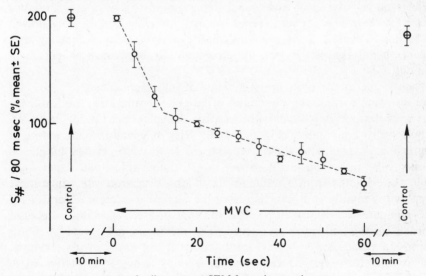

FIG. 8. (B) Mean values of spike counts ± SEM from six experiments.

also assumed to be representative of that of the muscle as a whole. These assumptions seem justified by the repeatability of the rate of action potential counts during a control series of brief maximal contractions both before and after fatigue took place. They are further supported by the finding that changes in both the spike count rate and the total integrated intramuscular EMG activity paralleled changes in the surface recording. Changes in the M wave recorded at the surface during submaximal stimulation also generated corresponding changes in the intramuscular record. Thus the behaviour of the fraction of the muscle being sampled appeared to be generally representative of the activity of the muscle as a whole.

The changes in the rate of overall muscle activation (spike counts) recorded during a sustained MVC were similar to those found to give the slowest rate of force loss and the best voluntary contraction match in the stimulation model. Changes in the absolute firing rate of individual motor units, however, could not be measured since interference from surrounding units generally prevented clear identification during the early part of contraction. Occasionally single units were seen during the first 10 s firing at between 25 and 35 Hz. In the later stages when individual units could easily be identified for short periods their rates rarely exceeded 10 and 15 Hz.

Frequency changes and motor unit types

Our results provide no evidence as to whether the changes in the overall rate of electrical activity result from equal changes in discharge rate of all motor neuron types or whether some types change rapidly and others only slowly. The relative benefit from changing motor neuron firing frequency may depend on differences between the electrical and mechanical properties of each unit type.

When single motor units are stimulated at high frequencies (80–100 Hz) there is a wide variation in their rates of fatigue. Despite this, the EMG for most units declines in parallel with the force (Kugelberg & Lindegren 1979), as it does for whole muscle (Fig. 6, p 140). This may well indicate failure of electrical propagation at these frequencies. If this also causes fatigue at natural firing rates one might expect that the high threshold, faster units would benefit most from a reduction in neuron discharge rate, since these fatigue most rapidly and may also have the highest maximum natural firing rate (Hannerz & Grimby 1979). But the rate of electrical failure depends critically on the activation frequency. The maximum firing rate of motor neurons in unfatigued voluntary contractions is not known with certainty. Few investigators report steady discharge rates above 35 Hz (Bigland & Lippold 1954, Clamann 1970, Milner-Brown et al 1973). During nerve stimulation at these lower frequencies failure of electrical propagation develops more slowly and is less pronounced. It is not clear, therefore, to what extent failure of electrical propagation would occur in voluntary contractions if the motor neuron discharge rates did not change.

Do changes in motor neuron firing rate cause or prevent force loss?

An argument has been put forward that a decline in activation frequency may reduce the rate of fatigue by avoiding the failure of electrical propagation which might otherwise occur. While the relative importance of this mechanism may depend on the natural firing frequency of the individual motor units, there may also be other functional consequences of such a process.

Motor neurons fire asynchronously with respect to each other. This random activity maintains the muscle series elasticity in a semi-stretched condition such that full tetanic tension may be achieved at frequencies substantially lower than are required during synchronous nerve or muscle stimulation (Rack & Westbury 1969, Lind & Petrofsky 1978). Thus the lower firing rates observed during maximal voluntary contractions are compatible with full muscle activation.

The loss of force seen during a maximal voluntary contraction sustained for 60 s is comparable to that produced if the frequencies of nerve stimulation of unfatigued muscle are reduced from 80 to 20 Hz (Fig. 4, p 139). A similar force loss during asynchronous voluntary activity would probably result if the average firing rate fell from 30 to 15 Hz. It is , therefore, tempting to suggest that much of the force loss observed during fatigue may result directly from the reduction in motor neuron firing rate. But this explanation is too simple. It overlooks the changes in the muscle contractile properties that occur simultaneously during fatigue.

The activation frequency required to tetanize a muscle is proportional to its speed. It is lower for slow than for fast muscles or motor units. During a 60 s MVC the overall muscle contractile speed (time to peak tension, relaxation rate, etc.) is reduced to about one-third of that in the unfatigued state (Edwards et al 1975). If the rate of contractile slowing corresponds to the rate of decline in motor neuron firing frequency the degree of tetanic fusion would not change and the muscle would remain functionally fully activated, even when firing rates were reduced to 10 or 15/s. No loss of force would, therefore, take place unless there were some failure at or beyond the transverse tubular system.

If the discharge rate did *not* decline the major functional disadvantage might be loss of motor control rather than of force *per se*. As a result of contractile slowing many motor neurons would now fire at supratetanic frequencies where rate modulation has no effect on tension (such behaviour would also be uneconomical). Maximum sensitivity requires that discharge rates never exceed the minimum required for full tetanic activation.

Control of frequency

To maximize force generation the firing frequency of each motor neuron should be regulated to activate the contractile system of its muscle fibres at intervals which are directly proportional to the twitch contraction time (Burke 1981). In unfatigued human muscles of mixed fibre composition there is commonly a 2–3-fold range of contraction times between the fast and slow motor unit types. This would require a corresponding range in maximum firing rates for fast and slow units if all are to be fully activated while still maintaining maximum control for force modulation. Furthermore, if this ideal relationship is to be maintained in fatigue the decline in motor neuron discharge rate must be geared to the rate of contractile slowing, which varies according to the type and intensity of exercise, the degree of ischaemia, etc. Does this process depend solely on the activation history of the motor neuron, or can the discharge rate be more precisely regulated by some feedback mechanism from the muscle itself?

Motor neurons can discharge at much higher rates than they normally do when activated voluntarily or by reflex pathways. Motor neuron discharge rates may be limited during prolonged activity by the duration of spike after-hyperpolarization, accommodation of excitability, mounting Renshaw inhibition, and various local inhibitory reflexes. An *increase* in firing frequency was observed by Hannerz & Grimby (1979) when the afferent inflow from the muscle was occluded during maximum voluntary efforts. During fatigue the normal inhibitory influence from Golgi tendon organs presumably declines with force loss, but concomitant fatigue of intrafusal fibres might reduce the excitatory response of muscle spindles to γ or β activation. In addition, free nerve endings within the muscle may inhibit motor neuron activity in response to muscle stretch. Thus, adequate machinery seems to exist for a feedback system signalling changes in contractile speed so that motor neuron firing rate may be appropriately modified. Whether or not such a regulatory process actually exists remains to be determined.

Conclusion

These experiments indicate that during sustained maximal isometric volun-tary contractions the muscle remains fully activated and force loss is not due to impairment of electrical propagation despite the reduction in electrical activity that may be observed. This reduced activity seems to result from a decline in motor neuron firing rate that matches the declining activation frequency required to tetanize the fatigued muscle as its contractile speed slows. Such a change in discharge would serve to maintain control of force modulation as well as to prevent failure of electrical propagation. Possibly the match between maximum motor neuron firing rate and changing muscle contractile properties is brought about, in part, by some reflex feedback mechanism.

Acknowledgements

I am grateful to the following colleagues who collaborated in various phases of these experiments: Professor R. H. T. Edwards and Drs G. P. Hosking, D. A. Jones, C. G. Kukulka, O. C. J. Lippold and J. J. Woods. This work was supported by US Public Health Service Grants NS 09960 and NS 14756 and by the US Muscular Dystrophy Association.

REFERENCES

Asmussen E 1979 Muscle fatigue. Med Sci Sports 11:313-321
Barcroft H, Millen JLE 1939 The blood flow through muscle during sustained contraction. J Physiol (Lond) 97:17-31
Bigland B, Lippold OCJ 1954 Motor unit activity in the voluntary contraction of human muscle. J Physiol (Lond) 125:322-335
Bigland-Ritchie B 1978 Factors contributing to quantitative surface electromyographic recording and how they are affected by fatigue. Am Rev Respir Dis 119:95-97
Bigland-Ritchie B, Edwards RHT, Jones DA 1977 Fatigue in sustained isometric voluntary contractions. Physiol Can 8:28
Bigland-Ritchie B, Jones DA, Hosking GP, Edwards RHT 1978 Central and peripheral fatigue in sustained maximum voluntary contractions of human quadriceps muscle. Clin Sci Mol Med 54:609-614
Bigland-Ritchie B, Jones DA, Woods JJ 1979 Excitation frequency and muscle fatigue: electrical responses during human voluntary and stimulated contractions. Exp Neurol 64:414-427
Bigland-Ritchie B, Lippold OCJ 1979 Changes in muscle activation during prolonged maximal voluntary contractions. J. Physiol (Lond) 292:14P-15P
Burke RE 1981 Motor units: anatomy, physiology and functional organization. In: Brooks VB (ed) Motor systems. Williams & Wilkins, Baltimore (Handb Physiol sect 1 The nervous system vol 4), in press
Clamann HP 1970 Activity of single motor units during isometric tension. Neurology 20:254-260
Clamann HP, Broecker KT 1979 Relation between force and fatigability of red and pale skeletal muscles in man. Am J Phys Med 58:70-85

Edwards RHT, Hill DK, Jones DA 1975 Metabolic changes associated with slowing of relaxation in fatigued mouse muscle. J Physiol (Lond) 251:287-301

Edwards RHT, Young A, Hosking GP, Jones DA 1977 Human skeletal muscle function: description of tests and normal values. Clin Sci Mol Med 52:283-290

Hannerz J, Grimby L 1979 The afferent influence on the voluntary firing range of individual motor units in man. Muscle Nerve 2:414-422

Humphreys PW, Lind AR 1963 The blood flow through active and inactive muscles of the forearm during sustained hand-grip contractions. J Physiol (Lond) 166:120-135

Ikai M, Yabe K, Ishii K 1967 Muskelkraft und muskuläre Ermüdung bei willkürlicher Anspannung und elektrischer Reizung des Muskels. Sportarzt Sportmedizin 5:197-211

Jones DA 1981 Fatigue due to changes beyond the neuromuscular junction. This volume, p 178-192

Jones DA, Bigland-Ritchie B, Edwards RHT 1979 Excitation frequency and muscle fatigue: mechanical responses during voluntary and stimulated contractions. Exp Neurol 64:401-413

Krnjević K, Miledi R 1958 Failure of neuromuscular propagation in rats. J Physiol (Lond) 140:440-461

Kugelberg E, Lindegren B 1979 Transmission and contraction fatigue of rat motor units in relation to succinate dehydrogenase activity of motor unit fibres. J Physiol (Lond) 288:285-300

Lind AR, Petrofsky JS 1978 Isometric tension from rotary stimulation of fast and slow cat muscles. Muscle Nerve 1:213-218

Merton PA 1954 Voluntary strength and fatigue. J Physiol (Lond) 128:553-564

Merton PA, Hill DK, Morton HB 1981 Indirect and direct stimulation of fatigued human muscle. This volume p 120-126

Milner-Brown HS, Stein HS, Yemm R 1973 Changes in firing rate of human motor units during linearly changing voluntary contractions. J Physiol (Lond) 230:371-390

Mosso A 1915 Fatigue, 3rd edn. Drummond M, Drummond WG (transl) Allen & Unwin, London

Naess K, Storm-Mathisen A 1955 Fatigue of sustained tetanic contractions. Acta Physiol Scand 34:351-366

Rack PMH, Westbury DR 1969 The effect of length and stimulus rate on tension in the isometric cat soleus muscle. J Physiol (Lond) 206:443-460

Reid C 1928 The mechanism of voluntary muscular fatigue. Q J Exp Psychol 19:17-42

Simonson E 1971 Physiology of work capacity and fatigue. Thomas, Springfield, Illinois, p 571

Stephens JA, Taylor A 1972 Fatigue of maintained voluntary muscle contraction in man. J Physiol (Lond) 220:1-18

Thesleff S 1959 Motor end-plate desensitization by repetitive nerve stimulation. J Physiol (Lond) 148:659-664

DISCUSSION

Clamann: So far in this symposium the discussion has focused on the fatigue properties of entire muscles. Muscles are composed of motor units which may differ widely in their fatigue properties; muscle fatigue is thus the aggregate of motor unit fatigue. I would like to illustrate some fatigue properties of different types of motor units, with emphasis on electrical fatigue properties that have not been much studied.

Single medial gastrocnemius motor axons were identified in finely divided ventral roots of cats anaesthetized with chloralose. Each motor unit was then stimulated via its motor axon, and mechanical properties including twitch and tetanic tension, 'sag', twitch contraction time and fusion frequency were measured. This allowed each unit to be characterized as type FF, FI, FR, or S according to the scheme of Burke (Burke et al 1973, Burke & Edgerton 1975). While we did not characterize the units histochemically, it has been shown that a good correspondence exists between the mechanical S, FR and FF unit types and the histochemical I, IIA and IIB types respectively (Burke & Edgerton 1975).

When a unit had been identified by its mechanical properties, we stimulated it continuously at 80 pulses per second (p.p.s.) to produce mechanical or electrical fatigue. The tetanic tension produced was recorded with a force transducer attached to the medial gastrocnemius muscle, and EMG activity was recorded between that pair of four bare wires, implanted at equal intervals into the belly of the muscle, which gave the largest amplitude signal. Each EMG waveform was full-wave rectified and its integral obtained with an electronic integrator. Motor units were stimulated at 80 p.p.s. because lower rates do not produce adequately fused tetani in fast units, and at higher rates the EMG waveforms overlap so that their integrals cannot be measured. All motor units were stimulated at the same rate so that direct comparisons of fatigue properties could be made.

Fig. 1 shows the fatigue properties of a fast and fatiguable unit (type FF or IIB). A shows the decline of the raw EMG waveform and force when the unit is stimulated at 80 p.p.s. Force declined by about 90% in three seconds (the horizontal time mark is five seconds), while the decline in EMG activity took about twice that time. B is an X–Y plot showing the relative decline of force against integrated EMG. The curve is traced left and downward from the right. Note that the initial fatigue effect was a force decline with very little change in IEMG. During this time the raw EMG waveform could be seen to decline in amplitude and increase in duration. C is an illustration of the fatigue index (Burke et al 1973). When 300 ms bursts of 40 p.p.s. stimuli were repeated once a second, force declined to zero while EMG was largely unaffected. Time mark in C is 15 seconds.

Fig. 2 shows the fatigue characteristics of an FR (IIA) unit. Both the EMG and force declined more slowly (Fig. 2A, time mark 5 s) so that 15–20 s were required to reduce force to 0. The X–Y plot shows that after an initial potentiation of the EMG integral, force and IEMG declined together. The IEMG potentiation occurred because the raw EMG waveforms increased in duration without much loss of amplitude in the first few seconds. Fig. 2C is again an illustration of the fatigue index, showing that force declined by about 15% during this two-minute test.

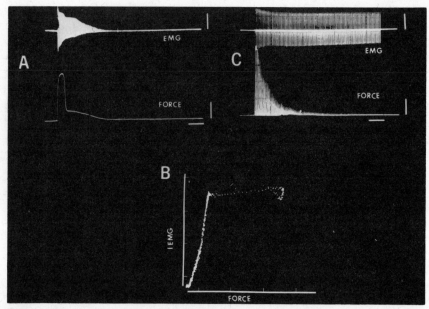

FIG. 1. (Clamann). The fatigue properties of a fast (type FF, or IIB) motor unit of the medial gastrocnemius muscle of the cat. (See text for details.)

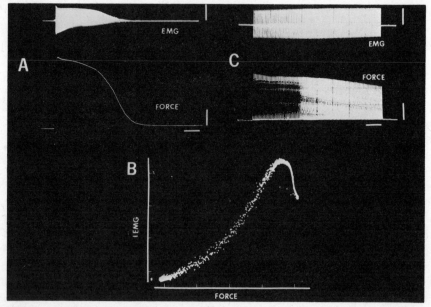

FIG. 2. (Clamann). The fatigue properties of an FR (IIA) motor unit of the medial gastrocnemius muscle of the cat. (See text for details.)

Fig. 3 shows the electrical and mechanical fatigue properties of an S (Type I) unit. The time mark is 15 seconds; EMG amplitude declined by about 50%, and force by about 15% in the 2½ minutes shown in A. In B it is seen that IEMG finally declined almost to zero with little concomitant force decline.

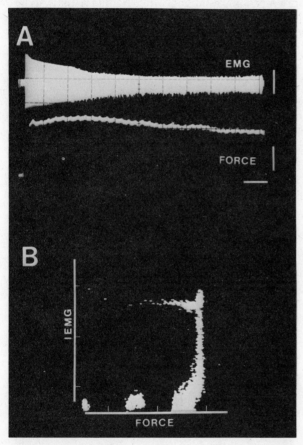

FIG. 3. (Clamann). The fatigue properties of an S (type I) motor unit of the medial gastrocnemius muscle of the cat. (See text for details.)

This is possible since the unit was being stimulated at far above its fusion frequency of about 25 p.p.s. The EMG waveform declined smoothly for about 30 seconds, and then began to show amplitude 'jitter', which is visible in the raw EMG record in A. Such 'jitter' always appeared after about 30 seconds of stimulation in S units, was often present in the later stages of fatigue of FR units, and was rarely seen in FF units. It is likely that 'jitter' takes more time to develop than the time required for a smooth decline to 0 of the EMG waveform. Hence it cannot be seen in FF unit fatigue. This 'jitter' is probably produced by the intermittent failure of electrical transmission in nerve branches or at the neuromuscular junction.

In summary, when motor units are stimulated at 80 p.p.s., type FF units initially show an almost pure mechanical fatigue, while S units initially show a pure electrical fatigue. In FR units, electrical and mechanical fatigue typically occur together. Type FF units show rapid (less than 10 seconds) loss of electrical and mechanical activity, while in S units fatigue takes minutes. In each of the three motor unit types, electrical and mechanical properties are seen to be well matched, with type S units showing the greatest mechanical and electrical fatigue resistance, and FF units showing the least.

Merton: May I show a slide to argue that electricity has everything to do with isometric fatigue and chemistry has nothing to do with it? Dr Ritchie showed us pictures of force plotted against time. Suppose we plot force against the number of impulses (say 3600, at 60 per second) as ordinate (Fig. 4). Then we repeat this but starting at 60 per second and decreasing the rate of stimulation progressively, down to 20 per second, in such a manner as

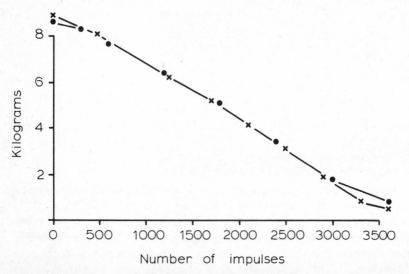

FIG. 4. (Merton). A comparison of the force in a tetanus at 60/s and in an 'artificial wisdom' run, as a function of the number of stimuli delivered. The artificial wisdom run consisted of 8 s at 60/s, 17 s at 45/s, 15 s at 30/s and 95 s at 20/s, a total of 3600 stimuli. It lasted 2 min 15 s. The 60/s tetani lasted 1 min and, thus, also contained 3600 stimuli. The curve plotted is the mean of two runs at 60/s, one made before the artificial wisdom run and one after. (Marsden et al 1976.)

to keep force up to the best possible amount. If we plot that second contraction in terms of the number of impulses, this line and the 60 per second line lie on top of each other. Thus, in this situation, the force depends solely on the number of shocks delivered. If you deliver them slowly enough you get, in fact, $2\frac{1}{4}$ times the tension–time out of the muscle. It is difficult to see how chemistry can govern here.

Edwards: How is this effect mediated? Is it through accumulation of potassium, depletion of sodium and so on? In other words, it may not be chemistry at the level of the contractile mechanism, but it may be 'chemistry' related to membrane ion exchanges.

Wilkie: And how can you be so sure about what you call 'chemistry' (whatever that means), without having any idea what the chemical changes are in the two different patterns of stimulation?

Merton: Is it possible, then, that the chemical changes after one minute of stimulation at 60 per second are the same as the chemical changes after 2.4 minutes at a varying frequency of stimulation?

Wilkie: It is perfectly possible. In our nuclear magnetic resonance (n.m.r.) experiments (Dawson et al 1978, 1980a,b) we deliberately varied the pattern of stimulation. We were stimulating intermittently, of course. We varied the 'duty cycle', stimulating for a second every 20 seconds, a second every minute, or five seconds every five minutes, so that we were varying the duty cycle in two different ways. What we found, and might well find in your experiment, for all I know, is that force and the chemical concentrations are correlated regardless of how the changes in both were brought about.

Merton: You are in general right, of course, and I am not really maintaining such an extreme position! But the experiment does show that, in this particular situation, if you want the best tension–time you must spread the motor impulses out.

Edwards: Assuming you can use integrated force–time as a guide to total metabolic cost, for the same integrated force–time would you expect the end-point chemistry to be similar?

Wilkie: No! (See Dawson et al 1978.)

Wiles: In an electrically stimulated contraction the energy turnover rate per unit force increases with decreasing tetanic fusion (Wiles 1980). Alternatively, the same force–time sustained by long contractions costs less energy than if sustained by several shorter contractions (Dawson et al 1978). Thus it is energetically advantageous to maintain force using a minimum number of muscle reactivations. In a voluntary contraction we cannot be certain of the degree of fusion of individual muscle fibre responses but one might speculate that the nervous system so adjusts the firing rate that, as slowing of relaxation occurs, optimum fusion is maintained. This ensures that the energy stores are used in the most economical way.

Edwards: I agree with that. We have to be careful, though, about claiming that altering the stimulation frequency by itself can alter the chemistry, if the force is the same in the two circumstances. Is that what you are postulating, Professor Wilkie?

Wilkie: I wasn't postulating anything! But Dr Merton's point that correlations such as we and others have shown don't make a distinction between a failure of excitation in a general sense and failure of the contractile mechanism is of course true. I imagine that one of the themes of this meeting is to try to disentangle the two, because both are involved in some way, and it's our major task to see in what way. I made a proposal which any theory has to satisfy, namely that however the muscle works, it doesn't work to the extent of ever letting the ATP level fall dangerously low.

Stephens: Peter Clamann showed motor unit EMG amplitude declining during fatiguing tetanic contractions (p 149). Brenda Ritchie, on the other hand, sees no change in the whole-muscle action potential when she stimulates the muscle nerve during fatigue from a maximum voluntary effort. If unit action potentials can be shown to decrease in amplitude on continuous stimulation in animal experiments, I would have expected the nerve-evoked, whole-muscle action potentials to change in amplitude during human fatiguing contractions, especially at the start of sustained maximum voluntary contractions, where motor unit firing frequency is high and the blood flow is occluded.

Clamann: I found that if I used a very different (slower) stimulus pattern, the one that Burke used for his fatigue index (Burke et al 1973), I did not see a decline in the EMG.

Bigland-Ritchie: That is my answer too. It all depends on the frequency. Voluntary contractions are very different from those generated by synchronous nerve stimulation. During nerve stimulation we also saw the EMG decline whenever the frequency was held constant, and above about 30/s. We had to reduce it from 60/s to 20/s to avoid a reduction in the evoked M wave. In fatigue of voluntary contractions, propagation failure, either at the neuromuscular junction or at the muscle surface membrane, does not appear to occur, probably because the steady-state, asynchronous motor neuron firing rates are lower. Maximum forces seem to be generated by discharge rates of only about 30s, which then decline rapidly to 10-15/s (our unpublished observations and average values from the literature; see Grimby et al, this volume p 157-165). In the rat diaphragm *in situ*, Krnjević & Miledi (1958) found failure of transmission with nerve stimulation at 50/s but none at 10/s. When classifying the fatigue properties of motor units, continuous stimulation at high frequencies (80-100/s) or intermittent bursts at 40/s may be appropriate, but their EMG responses then do not necessarily reflect their behaviour during voluntary contractions. We need to know more about the natural firing rates. However, even when stimulating at 80/s Dr Clamann showed an FF unit where the tension fell well before any failure of the EMG.

Clamann: We find that at a frequency of 30 per second, an FR unit takes over a minute for the EMG to show a significant decline.

Stephens: What about an FF unit?

Clamann: We haven't found one yet! Once we have fatigued a unit in our experiments, the most dramatic change is that in a subsequent test the time course of fatigue is very different. We have to allow 45–60 minutes for units to recover. I am only now trying different stimulus rates in the same unit, and those few experiments have yielded only a small number of FR and S units, but no type FF.

Stephens: This is a crucial point, because it helps to validate (or otherwise) the whole-muscle M waves. If in animal experiments and the best conditions, with blood flow intact, the FF unit action potentials decline, even when stimulated at the modest frequency of 40 per second and over a period of only 330 ms, there is an inconsistency between the two preparations. This is important because it is argued that, if no change in the size of the nerve-evoked muscle action potential is seen in a human experiment, there is no failure of neuromuscular transmission. An important part of that argument rests on how selective the EMG recording site is. It is important that we reconcile these two conflicting pieces of information.

Roussos: Dr Ritchie, you showed that the EMG was increasing at the beginning of voluntary contraction. You explained this by the spreading of the action potential. Did you actually measure the action potential of single motor units?

Bigland-Ritchie: No. We cannot generally follow the waveform of single units during the first 10 seconds of contraction.

Edwards: Of course, central and peripheral fatigue aren't mutually exclusive. Central fatigue certainly can occur, as some of you who study patients with muscle weakness know. It happens that a lot of the work has been done in well-motivated subjects. Another, perhaps more interesting point, is the relative motor control of different muscle groups. In our study with Brenda Ritchie (Bigland-Ritchie et al 1978) central fatigue was demonstrated in the quadriceps muscle. I gather that it is more difficult to demonstrate central fatigue in sustained contraction of the adductor pollicis muscle. It is an interesting question whether there is a difference, related to motor unit size or afferent control of those motor units, between the large quadriceps muscle and a small hand muscle.

Bigland-Ritchie: That may well be true, and that is why I emphasized the problem of motivation. I have defined central fatigue here as something that could *never* be overcome voluntarily, even briefly; for example, as changes in lower motor neuron excitability. The separation of changes in voluntary motivation (conscious or unconscious) during a contraction, from a built-in, irreversible fatigue mechanism is very difficult.

Edwards: That is an important point to establish, because so far the definition of central fatigue has been based on a comparison between the forces achieved with electrical stimulation and with a voluntary effort. It may be that voluntary effort can be changed just by changing the recruitment circumstances or by a brief 'super' effort, producing recruitment beyond what is achieved in a normal sustained contraction. Are we going to include under the term 'central fatigue' the reduction in motor unit firing frequency that occurs to optimize muscle contractile function?

Bigland-Ritchie: I would not include it. Following your lead, we define fatigue as a loss of *force*. The reduction in motor neuron firing rate is certainly a change which takes place with prolonged activity. These changes, which are *not* necessarily associated with loss of force, are sometimes referred to as fatigue and cause confusion in the literature. A new and clearer set of definitions is needed.

Edwards: If we keep to loss of force as the definition, then during the period when there is very little loss of force there is, undoubtedly, a fall in the firing frequency.

Bigland-Ritchie: Yes.

REFERENCES

Bigland-Ritchie B, Jones DA, Hosking GP, Edwards RHT 1978 Central and peripheral fatigue in sustained maximum voluntary contractions of human quadriceps muscle. Clin Sci Mol Med 54:609-614

Burke RE, Edgerton VR 1975 Motor unit properties and selective involvement in movement. Exercise Sport Sci 3:31-81

Burke RE, Levine DN, Tsairis P, Zajac FE 1973 Physiological types and histochemical profiles in motor units of the cat gastrocnemius. J Physiol (Lond) 234:723-748

Dawson MJ, Gadian DG, Wilkie DR 1978 Muscular fatigue investigated by phosphorus nuclear magnetic resonance. Nature (Lond) 274:861-866

Dawson MJ, Gadian DG, Wilkie DR 1980a Mechanical relaxation rate and metabolism studied in fatiguing muscle by phosphorus nuclear magnetic resonance. J Physiol (Lond) 299:465-484

Dawson MJ, Gadian DG, Wilkie DR 1980b Studies of the biochemistry of contracting and relaxing muscle by the use of ^{31}P n.m.r. in conjunction with other techniques. Proc R Soc Lond B Biol Sci 289:445-455

Krnjević K, Miledi R 1958 Failure of neuromuscular propagation in rats. J Physiol (Lond) 140:440-461

Marsden CD, Meadows JC, Merton PA 1976 Fatigue in human muscle in relation to the number and frequency of motor impulses. J Physiol (Lond) 258:94P-95P

Wiles CM 1980 The determinants of relaxation rate of human muscle *in vivo*. PhD thesis, University of London

Firing properties of single human motor units on maintained maximal voluntary effort

LENNART GRIMBY, JAN HANNERZ, JÖRGEN BORG and BJÖRN HEDMAN

Department of Neurology, Karolinska sjukhuset, S-104 01 Stockholm, Sweden

Abstract The discharge properties on maintained maximal voluntary effort, axonal conduction velocity (a.c.v.) and contraction time (c.t.) of single human motor units were studied. Electromyographic (EMG) techniques were used and sufficient selectivity was obtained after repeated lesions to the terminal nerve twigs and consequent collateral sprouting, or by blocking the main muscle nerve in subjects with an accessory nerve supplying just one or a few units, or using high impedance wire electrodes. Most units with a.c.v. above 45 m/s and c.t. below 50 ms had maximal voluntary firing rates of about 50 Hz. On prolonged maximal effort, however, their rates rapidly decreased and after some seconds to a minute they ceased to respond tonically. As long as their motoneurons fired their EMG potentials were mainly intact and their twitch tension was significant. Most units with a.c.v. below 40 m/s and c.t. above 60 ms fired initially at 30 Hz, decreased slowly in firing rate and continued at 20 Hz for some minutes. We conclude that there is a central fatigue of units with short c.t. and high a.c.v. which protects their peripheral regions from severe exhaustion, but that there is no significant central fatigue of units with long c.t. and low a.c.v. We emphasize, however, that most units are intermediate in their properties.

Peripheral fatigue of single motor units—that is, failure of their peripheral electrical propagation or contractile mechanisms—has been extensively studied, mainly in animals but recently also in man (Garnett et al 1979). Central fatigue of single motor units—insufficient activation of the motoneuron—has been studied much less, partly because of the complexity of the central nervous system and partly because of technical difficulties. We have, however, been able to study central fatigue of single anterior tibial motor units and short extensor motor units of the big toe in man. This review is based mainly on papers by Hannerz (1974), Grimby & Hannerz (1977), Borg et al (1978) and Grimby et al (1979, 1981).

1981 Human muscle fatigue: physiological mechanisms. Pitman Medical, London (Ciba Foundation symposium 82) p 157-177

Central fatigue

On supra-maximal electrical tetanization of the peroneal nerve, 50 Hz was necessary for maximum dorsiflexion tension of the big toe as well as maximum dorsiflexion tension of the foot. Stimulation at 100 Hz did not result in a higher tension but caused a more rapid increase in tension; stimulation at 40 Hz resulted in about 90% of maximum tension. On continuous tetanization at 50 Hz, tension remained relatively high for about 20 seconds but thereafter tension rapidly decreased to 30–20% of maximum (curve a in Fig. 1). There was a simultaneous decrease in the electromyogram (EMG), indicating insufficiency of peripheral electrical propagation.

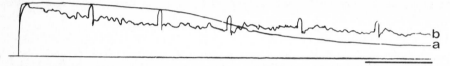

FIG. 1. Dorsiflexion tension of the big toe obtained by supra-maximal electrical tetanization of the peroneal nerve at 50 Hz (curve a) and by maximum voluntary effort with regularly superimposed electrical tetanizations (curve b). Time bar, 10 s.

Maximum tension could be attained but not maintained voluntarily (curve b in Fig. 1). The relative roles of central and peripheral fatigue in voluntarily maintained maximum contraction were determined by superimposing supra-maximal electrical tetanization of the peroneal nerve at 50 Hz. On prolonged maximal effort there was a decrease in the voluntary tension as well as the tension evoked by superimposed tetanization. Voluntarily maintained tension, however, decreased more rapidly; that is, there was central fatigue. After one or a few minutes a steady state was achieved when about one-third of maximum tension could be maintained voluntarily, one-third was lost by central fatigue and one-third lost by peripheral fatigue. The collapse of the peripheral conduction described for prolonged electrical tetanization was not present on prolonged maximum effort, indicating that central fatigue protects peripheral conduction from breakdown, as previously described by Bigland-Ritchie et al (1979).

The regularly superimposed electrical tetanizations informed the subject continuously that there were tension reserves in spite of the subjective impression of maximal contraction. Knowing this, subjects were capable of improving their capacity to maintain tension voluntarily over the experimental period of six months. In parallel with more intense voluntary use, electrically evoked tension decreased at a higher rate; that is, the relative importance of central fatigue decreased on training. Some central fatigue, however, always remained.

It must be emphasized that the subject at any time could reach the electrically evoked maximum tension voluntarily by first decreasing his voluntary drive and then making a maximal acceleration, but only for a few seconds. Thereafter, voluntarily maintained tension tended to fall below that expected if no acceleration were made.

The finding of central fatigue agrees with findings by Reid (1928), Ikai et al (1967) and others, but seems contrary to findings by Merton (1954) and Bigland-Ritchie et al (1979). It may be possible to overcome central fatigue by extraordinary motivation and training. Anyhow, maximal tension is so little accessible that in everyday life it is used only phasically.

Techniques for studying central fatigue of single motor units

We have studied central fatigue of single motor units in the short extensor of the big toe and the anterior tibial muscle on maximum voluntary effort so prolonged that severe fatigue developed. Electromyographic techniques were used. Supra-maximal electrical stimulation of the peroneal nerve was superimposed when the test unit no longer responded voluntarily, to differentiate between central fatigue and peripheral conductive fatigue. The study concentrated on the extreme types of motor units: those with high axonal conduction velocity and short contraction time, and those with low axonal conduction velocity and long contraction time.

Electromyographic recordings of sufficient selectivity were obtained in the short extensor muscle of the big toe in three ways. (1) The number of motor units was reduced by lesions to the terminal nerve twigs and muscle fibres. Subsequently, muscle fibre density within the remaining motor units increased by collateral sprouting. After repeated lesions over several months very high selectivity of EMG recordings could be attained. (2) The main deep peroneal nerve was blocked with lidocain in subjects with an accessory deep peroneal nerve (Lambert 1969) innervating just a few motor units. (3) Using high impedance wire electrodes, as described by Hannerz (1974), and considerable patience, we obtained sufficiently selective recordings in unprepared muscles. This third technique was the only one used on the anterior tibial muscle.

Technique 1 has the disadvantage that not only may the peripheral conductive and contractile mechanisms be abnormal but also the firing properties of the motoneuron, since the inflow from muscle receptors, essential for maintaining motoneuron firing (Hannerz & Grimby 1979), may be changed by the preparation. Also, the after-hyperpolarization of the motoneuron may be changed (Kuno et al 1974), its synapses may be partly rejected (Blinzinger & Kreutzberg 1968), and its firing properties may be disturbed (Eccles et al 1958) by lesions in its periphery. With technique 2 the proprioceptive inflow is disturbed but the other sources of error are eliminated. With technique 3 none of the sources of error mentioned are present but the lower selectivity does not allow systematic identification of single motor unit potentials on supra-maximal electrical stimulation of the peroneal nerve—that is, systematic differentiation between central fatigue and peripheral conductive fatigue. Similar findings obtained with different techniques should, however, be conclusive, since the risks of error are different.

The voluntary discharge properties on fatigue were studied with techniques 1 and 2 in prepared toe muscles and with technique 3 in the intact anterior tibial muscle.

The axonal conduction velocity (a.c.v.) and the contraction time (c.t.) were studied only in the short toe extensor muscle, using techniques 1 and 2. The conduction velocity of the single motor unit was calculated from the difference in latency of its EMG potential on proximal and distal nerve stimulation, as described by Borg et al (1978). The contraction time was calculated by averaging the increase in force related to its EMG potential in maintained voluntary contraction, as described by Milner-Brown et al (1973), or by stimulation of its intramuscular nerve twigs, as described by Taylor & Stephens (1976).

Firing rates of single motor units on maximum voluntary tension

When maximum tension (defined as the tension on supra-maximal tetanization at 50 Hz) was maintained in voluntary contraction the firing rates of the single motor units could be differentiated, as illustrated in Fig. 2. Motor units

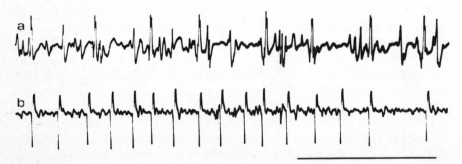

FIG. 2. Firing rate of two anterior tibial motor units (a and b respectively) on maximum voluntary effort resulting in a tension of 230 N. Time bar, 100 ms. The motor unit in (a) attains 7 discharges in 250 ms, i.e. 28 Hz. The motor unit in (b), on the other hand, attains 15 discharges in 250 ms, i.e. 60 Hz.

with an a.c.v. above 45 m/s and a c.t. below 50 ms had a maximum firing rate of about 50 Hz. This agrees with the finding that 50 Hz was necessary for maximum tension on electrical tetanization. Motor units with an a.c.v. below 40 m/s and a c.t. above 80 ms fired at only 30 Hz on maximum tension. This is in accordance with the fact that the firing interval necessary for complete fusion increases with the contraction time. It must be emphasized, however, that all motor units fired at higher rates at the very beginning of the contraction when tension was not yet built up. This might be expected, since very short firing intervals are needed for a rapid build-up of tension.

Firing properties of single motor units in fatigue

On prolonged maximum voluntary effort the firing rates of the single motor units decreased successively. The decrease was more rapid for motor units with high initial firing rates than for units with low rates, so that the differences in firing rate were eliminated.

Each motor unit had its threshold tension and its minimum rate below which it did not fire repeatedly. The threshold tension and the minimum rate were related to the a.c.v. and the c.t. of the single motor units. Motor units with an a.c.v. above 45 m/s and a c.t. below 50 ms had a high threshold and a minimum rate above 20 Hz. Motor units with an a.c.v. below 40 m/s and a c.t. above 80 ms has a low threshold and usually a minimum rate below 10 Hz. When, on fatigue, the voluntary tension and firing rate decreased below the critical values for a particular motor unit, it ceased to participate tonically.

On maintained maximum voluntary effort the first motor units ceased to fire tonically when 90% of the remaining maximum tension of the muscle (defined as the tension evoked by superimposed electrical tetanization at 50 Hz) and the rate of 30 Hz could no longer be maintained. This usually happened within 10 seconds of maximum voluntary effort. In untrained subjects the electromyographic potential of these early fatigued motor units was also mainly intact when their motoneurons were ceasing to respond tonically. Their twitch tension was also partly retained as long as their motoneurons fired. Thus, their muscle fibres were protected from excessive exhaustion on maintained voluntary contraction by central mechanisms. Studies are in progress with the aim of determining whether neuromuscular transmission becomes a limiting factor (cf. Stephens & Taylor 1972) when central fatigue is decreased by training.

It must be stressed, however, that high threshold motor units lost only their capacity to respond tonically to voluntary drive and retained their capacity to respond phasically on fatigue. Consequently, in long series of phasic voluntary contractions, such as in alternating movements, their peripheral structures are not protected from exhaustion as they are in prolonged tonic voluntary contraction. Thus, different structures are the limiting factors in different types of voluntary contraction.

On the other hand, firing rates of 15–20 Hz could be maintained for some minutes: motor units with lower minimum rates never ceased to fire tonically and muscle fibres showing complete fusion at such low rates were thus not protected from fatigue by central mechanisms.

In the experiment illustrated in Fig. 3 two single motor units were within

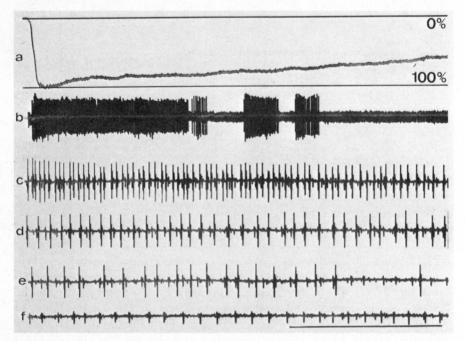

FIG. 3. Maintained maximum voluntary effort. (a) Dorsiflexion tension of the big toe. (b) EMG of one fatiguable motor unit (large amplitude potential) and one non-fatiguable (small potential) at the same film speed as the tension curve. (c) EMG at maximum tension at fast film speed. (d) EMG after about 5 s at fast film speed. (e) EMG after about 15 s at fast film speed. (f) EMG after about 25 s at fast film speed. Time bar a–b, 10 s; c–f, 500 ms. Further description in text.

the recording range of the electromyographic electrode. One of them had a high amplitude potential, the other a small amplitude. The motor unit with the large potential responded tonically to maintained maximum voluntary effort for only about 10 seconds. Thereafter it fired phasically a few times. After about 20 seconds it remained inactive in maintained maximum contraction (Fig. 3b). During the first second it fired at about 50 Hz (c). During the fifth second its firing rate was about 30 Hz (d). During the 15th second its firing rate decreased below 20 Hz, its minimum rate, and it ceased firing (e). The motor unit with the small potential, on the other hand, could be driven continuously for minutes at 20 Hz (f).

Mechanisms of central fatigue

The differences in the endurance of single motor units on voluntary drive can partly be explained by a general decrease in the efficiency of the drive, primarily affecting motor units with high thresholds (those with a high a.c.v. and a short c.t.). However, there were also signs of a selective increase in the threshold of such motor units. During prolonged strong voluntary contraction their firing rate decreased also when low thresholds units maintained or even increased their firing rate. Further, during fatigue of voluntary contraction, the threshold tension of high threshold units increased in relation to the remaining maximum tension of the muscle. Finally, as illustrated in Fig. 4,

FIG. 4. Dorsiflexion tension of the big toe (upper trace) and EMG of a fatiguable motor unit (lower trace) on voluntary effort increasing to maximum. 100% denotes the tension evoked by supra-maximal electrical tetanization at 50 Hz immediately after the experiment. Time bar, 1 s. Further description in text.

certain high threshold units fired initially at a moderately high tension but after some seconds they ceased to fire in spite of continually increasing tension, etc., until they finally fired only at a tension level too high to be maintained; that is, they fired mainly phasically. Low threshold units, with low a.c.v. and long c.t., on the other hand, did not behave in this way. Once recruited they continued to fire as long as their critical tension was maintained.

Relation between central fatigue, axonal conduction velocity and contraction time of single motor units

Fig. 5 summarizes the findings in the short toe extensor muscle. Most of the motor units with an a.c.v. above 45 m/s and a c.t. below 50 ms discharged

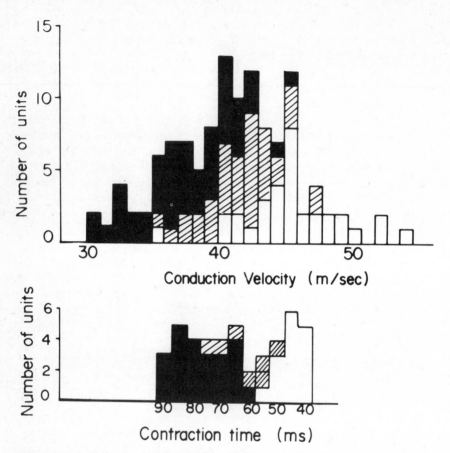

FIG. 5. Relation between axonal conduction velocity, contraction time and voluntary discharge properties. Open squares denote motor units which had a maximum rate above 40 Hz, a minimum rate above 20 Hz and which fired intermittently on maximum voluntary effort. Filled squares denote motor units which had a maximum rate of about 30 Hz, a minimum rate of about 10 Hz and which fired continuously at 15–20 Hz for minutes on maximum voluntary effort. Hatched squares denote intermediates.

only for limited periods of time. As mentioned above, they at least partially retained EMG potentials and twitch tension when their motoneurons were ceasing to respond. We conclude that there are fast twitch motor units whose muscle fibres are protected from severe fatigue by central mechanisms.

On the other hand, most motor units with an a.c.v. below 40 m/s and a c.t. above 80 ms continued to fire at 15-20 Hz. This rate should be sufficient for full fusion, at least in the fatigued muscle. We conclude that there are slow twitch motor units whose muscle fibres are not open protected from fatigue by central mechanisms.

We emphasize, finally, that a considerable number of the motor units had intermediate properties.

Acknowledgements

These investigations were supported by grants from the Swedish Research Council (project no. B81-14X-04749-068).

REFERENCES

Bigland-Ritchie B, Jones DA, Woods JJ 1979 Excitation frequency and muscle fatigue: electrical responses during human voluntary and stimulated contractions. Exp Neurol 64:414-427

Blinzinger K, Kreutzberg G 1968 Displacement of synaptic terminals from regenerating motoneurons by microglial cells. Z Zellforsch Mikrosk Anat 85:145-157

Borg J, Grimby L, Hannerz J 1978 Axonal conduction velocity and voluntary discharge properties of individual short toe extensor motor units in man. J Physiol (Lond) 277:143-152

Eccles JC, Libet A, Young RR 1958 The behaviour of chromatolysed motoneurones studied by intracellular recording. J Physiol (Lond) 143:11-40

Garnett RAF, O'Donovan MJ, Stephens JA, Taylor A 1978 Motor unit organization of human medial gastrocnemius. J Physiol (Lond) 287:33-43

Grimby L, Hannerz J 1977 Firing rate and recruitment order of toe extensor motor units in different modes of voluntary contraction. J Physiol (Lond) 264:865-879

Grimby L, Hannerz J, Hedman B 1979 Contraction time and voluntary discharge properties of individual short toe extensor motor units in man. J Physiol (Lond) 289:191-201

Grimby L, Hannerz J, Hedman B 1981 Fatigue and voluntary discharge properties of single motor units in man. J Physiol (Lond), in press

Hannerz J 1974 Discharge properties of motor units in relation to recruitment order in voluntary contraction. Acta Physiol Scand 91:374-384

Hannerz J, Grimby L 1979 The afferent influence on the voluntary firing range of individual motor units in man. Muscle Nerve 2:414-422

Ikai M, Yabe K, Ishii K 1967 Muskelkraft und muskuläre Ermüdung bei willkürlicher Anspannung und elektrischer Reizung des Muskels. Sportarzt Sportmedizin 5:197-204

Kuno M, Miyata Y, Muñoz-Martinez EJ 1974 Differential reaction of fast and slow alpha-montoneurones to axotomy. J Physiol (Lond) 240:725-739

Lambert EH 1969 The accessory deep peroneal nerve. A common variation in innervation of extensor digitorum brevis. Neurology 19:1169-1176

Merton PA 1954 Voluntary strength and fatigue. J Physiol (Lond) 123:553-564

Milner-Brown HS, Stein RB, Yemm R 1973 The contractile properties of human motor units during voluntary isometric contractions. J Physiol (Lond) 228:285-306

Reid C 1928 The mechanism of voluntary muscular fatigue. Q J Exp Physiol 19:17-42

Stephens JA, Taylor A 1972 Fatigue of maintained voluntary muscle contraction in man. J Physiol (Lond) 229:1-18

Taylor A, Stephens JA 1976 Study of human motor unit contractions by controlled intramuscular microstimulation. Brain Res 117:331-335

DISCUSSION

Merton: I would like to congratulate Dr Grimby on his extremely interesting results. Marsden, Meadows and I studied aberrant units in the adductor pollicis supplied by a communication between the median and ulnar nerves in the forearm (Marsden et al 1971) (Fig. 1). Some of our comparative figures

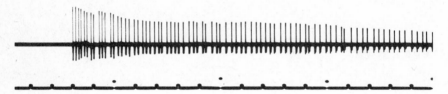

FIG. 1. (Merton). Action potentials from an aberrant median-supplied unit in the adductor pollicis at the start of a maximal voluntary effort. The ulnar nerve was blocked at the elbow. The peak frequency is 100 per second. A faster and more irregular unit is shown in Marsden et al (1971). Time markers at 1/10 second intervals with every fifth and tenth marker accentuated.

may be of interest. The maximum instantaneous rate of discharge that we saw was 200 per second; several units fired at over 150/s. We found that 50/s or 60/s is insufficient for maximum tension in the adductor pollicis; you need about 140/s in an electrically stimulated contraction, which matches the frequencies of the single units. The maximum rate of rise in a twitch in adductor pollicis from electrical stimulation is not achieved until 250/s, but we saw no single units firing faster than 200/s. We agree that they slow progressively, but we don't find that the final rates after a minute of maximum contraction are all the same. Some went down to about 15/s; others stayed up at about 25/s after a minute, in our study.

Grimby: Except for the first few seconds we found no difference in firing rate between motor units that were firing tonically. The motor units that were ceasing to fire tonically had a lower rate, of course. I should emphasize that our maximum rates were obtained on maintained maximum voluntary tension. During the stage of building up tension, much higher rates were found.

Merton: Yes; one doesn't keep up these rates for long. That is not surprising, because in an electrically excited tetanus lasting a second the largest peak tension in the adductor pollicis is reached within the first half-second with a rate of stimulation of about 140/s; but the tension immediately starts to come down and you may get a better end-result by switching to 100/s in the middle. So even in a tetanus of one second it may pay to slow for the second half; and the motor unit always seems to do the same in a voluntary contraction. It would be worse than useless to keep up these rates for long.

Edwards: It is a considerable technical achievement to record the firing frequency in maximal contraction conditions. In hearing about the frequency necessary to initiate maximal rates of force generation, we are essentially hearing about the beginnings of what might be ballistic (sudden) contractions. Others, particularly Desmedt & Godaux (1977), have recorded high frequencies, with ballistic contractions.

Bigland-Ritchie: I agree with Dr Grimby that if you decrease the effort a little and then restore it, you see a ballistic response and the firing frequencies clearly go up.

Grimby: During maintained voluntary effort the maximal firing rates can be restored if the voluntary drive is decreased from maximum and then rapidly increased.

Bigland-Ritchie: Did I understand you to say that during fatigue you see a fairly dramatic increase in the recruitment threshold?

Grimby: Yes, for at least some of the motor units with high axonal conduction velocity and short contraction time.

Bigland-Ritchie: If that is so, at submaximal forces one would expect the low threshold units to compensate by firing *faster* than they otherwise would.

Stephens: In Dr Merton's record (Fig. 1, p 166) one can see gaps in what is otherwise a regular spike train whose mean inter-spike interval is progressively lengthening. It is as if from time to time there are spikes missing from the train. Has Dr Grimby seen this, and how often have you seen it, Dr Merton? Are these gaps in the record evidence of neuromuscular transmission failure?

Merton: If they were, you would have to suppose that all the neuromuscular junctions in the motor unit fail synchronously.

Stephens: Not necessarily. To decide what is happening we need to know how many muscle fibres from the motor unit are contributing significantly to the recorded motor unit action potential. We should also be careful about interpreting this sort of record, because it is not taken during a normal maximum voluntary contraction. The force is not maximal. The muscle is relaxed, and so the occlusion of muscle blood flow that normally accompanies a maximum contraction is absent.

Grimby: At high voluntary firing rates there are 'gaps', suggesting failure either of the motor neuron discharge or of transmission in a terminal nerve twig. It is very difficult to superimpose an electrical stimulus in these 'gaps'. We have superimposed electrical stimulation each second, to minimize the risk of confusion between peripheral and central fatigue. But we have not done it when the motor unit is still firing at high rates.

Stephens: It is with a combination of high firing rates and occluded blood supply that you are likely to see failure of neuromuscular transmission. This will be in the first 20–30 seconds of a maximum effort. Dr Merton's record looks as if part of the motor unit action potential is breaking up.

Bigland-Ritchie: The record is compatible with that interpretation, but many of the apparent 'gaps' are preceded by an unusually short inter-spike interval. There may not be any missing.

Stephens: Peter Clamann once constructed histograms of motor unit firing intervals. Would Dr Merton's record give a normal histogram?

Clamann: Years ago, when I constructed motor unit inter-spike interval histograms (Clamann 1969), I was interested in the steady-state properties and threw away the interesting initial portions of some records. Motor units begin to discharge irregularly at some frequency and after about 10 seconds settle down to a lower discharge rate which is very regular. *That* is when I looked at them and obtained Gaussian inter-spike interval histograms. This record of Dr Merton's would give you a histogram with several peaks.

Stephens: Wouldn't this be exactly what you would get if the muscle fibres being recorded from dropped out from time to time, due to failure of neuromuscular transmission?

Clamann: Yes, but I have seen something that could provide a different explanation. When you stimulate extraocular motor neurons intracellularly and record EMG and force of the motor unit, you can see that type of pattern. Dr Stephen Goldberg and I (unpublished) tried to fatigue them and never could, because the units would begin to discharge intermittently, producing such gaps and allowing some fatigue recovery. Perhaps something is changing the sensitivity of the motor neuron. Then, of course, the whole unit would drop out and not just a few fibres.

Merton: Is Dr Stephens seriously suggesting that a gap arises because all the muscle fibres in a motor unit suddenly, early in the contraction, develop synchronous neuromuscular block for 5–10 ms, and that this is affecting all muscle fibres within reach of the electrode?

Stephens: How many muscle fibres lie within the reach of the recording electrode? That is the crux of the matter. If the record is dominated by the action potential of one muscle fibre close to the needle tip, then the gap in the record is the result of that one fibre dropping out. What the other fibres in the motor unit are doing, I don't know. It is a matter of interpreting the records correctly. If you were using conventional monopolar recording, the result would be different from bipolar recording.

Merton: We used, in fact, bipolar needle electrodes.

Stephens: A bipolar needle would make the recording area even smaller. The pick-up from a bipolar needle declines almost to zero over 100 μm (Stalberg & Trontelj 1979). On average there will be only one muscle fibre from a given motor unit within 100 μm of the needle tip. The unit electrical record can quite easily be dominated by the electrical activity generated by one muscle fibre.

Merton: You cannot really be suggesting that we were always recording from a single muscle fibre? The most interesting thing about such records, in fact, is that they show that a motor unit never fires regularly in the intact subject.

Stephens: In your opinion, Dr Grimby, how many muscle fibres contribute significantly to the unit electrical signal recorded by a bipolar EMG needle?

Grimby: A number. Of course, I can adjust the location of the needle so that the potential is dominated by one muscle fibre, but to retain it there is difficult.

Lippold: On the other hand, the obvious thing about Dr Merton's record is that the amplitude of the motor unit action potentials declines with time. You could interpret that as neuromuscular block, if you really wanted to. Individual muscle fibres drop out of the contraction as it proceeds, because of transmission failure.

Stephens: The irregularity of firing seen in Dr Merton's record is not normal. You don't see these occasional long gaps between motor unit spikes in a normal non-fatiguing contraction.

Merton: The irregularity that one sees in Fig. 1 is normal in a maximal contraction.

Grimby: The irregularity is least at a medium firing rate. When the rate is decreased, the irregularity increases. When the rate is increased to maximum, the irregularity increases also.

Merton: You agree with our impression that you never can get a single unit discharge which is quite regular?

Grimby: Yes.

Edwards: Can we move on to the factors that control the firing frequency? An experiment that does not appear to have been followed up, and perhaps should be, is that of Freyshuss & Knutsson (1971), who recorded the electroneurogram from the ulnar nerve. It is not possible to record motor nerve action potentials from a mixed nerve since the traffic is two-way and most fibres of a peripheral nerve are afferent rather than efferent. To record the motor nerve firing frequency they had to secure complete curarization. During a maximal voluntary effort, in which there was no contraction because of the curarization, the maximum frequency achieved was only about 30 impulses per second. Also, the firing frequency fell off rapidly. This raises the question of how important peripheral afferent information is in (1) pushing up the firing frequency, and (2) sustaining the firing frequency. Dr Grimby has studied motor unit firing frequency in relation to altered afferent information, with partial nerve block. What do you think?

Grimby: We studied single anterior tibial and short toe extensor motor units and blocked the sciatic nerve with ischaemia until only 5% of the tension remained in the muscle; that is, until most large afferent fibres and most α-efferents should have been blocked. The remaining motor units fired above 50 Hz for 30 seconds or a minute and their firing rate did not decline (Hannerz & Grimby 1979).

Edwards: Can we be clear on one other thing? You have shown that the different thresholds for recruitment of motor units are well separated in normal circumstances, during a gradual, hierarchical recruitment pattern, and that this recruitment pattern is lost, or the thresholds are brought closely together, when you alter afferent stimuli. Is that right?

Grimby: The recruitment pattern is not lost, but it is changed (Grimby & Hannerz 1976). I must emphasize that this is in advanced blockade of the afferent input. The recruitment mechanism seems to be resistant to loss of input until a certain limit is reached.

Edwards: Do you suggest that the hierarchical pattern of recruitment depends on the afferent information?

Grimby: Yes, partially.

Stephens: That is right. It depends also on the *sort* of afferent input. The responsiveness of different motor neurons to muscle afferent input is graded in just the same way as they are recruited in a voluntary contraction (Buller et al 1980), whereas their response to cutaneous stimuli is quite the reverse (Garnett & Stephens 1980). Skin stimulation facilitates activity in normally high threshold units and inhibits the activity of lower threshold units (Datta & Stephens 1979). By choosing suitable pairs you can make motor unit recruitment go in a reverse order. You can reverse it completely, in fact (Stephens et al 1978, Buller et al 1978, Garnett & Stephens 1978, 1981).

Clamann: Yes. Using a stimulus paradigm in which conditioning shocks to the sural nerve inhibited (that is, reduced the size by 20–50%) a subsequent monosynaptic reflex in the medial gastrocnemius muscle, we could show that motor units contributing to this reflex were recruited in inverse order of size (Clamann & Robinson 1980). That is, motor units with fast conducting axons were recruited first, and recruitment continued with units of progressively decreasing conduction velocities. The smallest units were recruited last.

Edwards: May we return to the specific question of the control of firing frequency during a sustained maximal voluntary contraction, and how it is that firing frequency is sufficiently high initially to give the figures that people record, and it falls off with time? We know that a ballistic (or sudden) contraction can occur for a short period, but something must rapidly reinforce that frequency. Do we have the necessary neural afferent information to support that? If so, what neural afferents are we talking about? Later, when the firing frequency has fallen and force has also fallen appreciably, are we seeing a failure of the same mechanism, or are we recognizing that the sensation is very different when you are producing less force? I am talking now specifically about muscle afferents rather than cutaneous ones.

Bigland-Ritchie: If one accepts the argument that with fatigue a reduction in the *range* of motor neuron firing frequencies is important for motor control, one must think in terms of a control mechanism that will match this with the slowing of the muscle contractile properties. This can change at different rates, and also under other conditions, which do not necessarily involve fatigue, as with changes of temperature.

Edwards: Can we bring in the question of curarization, which induces neuromuscular block? We have talked about the phenomenon in the completely curarized state, but some interesting effects have been experienced in the partially curarized state. It is probably worth expanding the discussion here to including the perception of fatigue, of effort and of the loss of force. Moran Campbell has had most experience of being both completely curarized (Campbell et al 1969) and also partially curarized (Campbell et al 1977).

Campbell: I am diffident about entering this territory, because the insistence that fatigue is defined for operational purposes in terms of force has clearly been valuable for the progress of neurophysiology. However, there is something which I hope to have explained. My point is that the sense of fatigue is not necessarily a sense of weakness. The basis of this contention comes from my own experience that being paralysed by curare is not the same as being paralysed by prolonged contraction or ischaemic contraction. The first time I was partially curarized, I was amazed that the limb that I was told to move 'was unwilling to move'. There was nothing like the sense of 'if only I could try harder, it would move', which I knew from previous experiments on myself, and was hinted at by references to the subjects being urged to 'try harder' in the experiments described earlier. Perhaps this experience is entirely confined to me or is too subjective to be trusted, yet there seems to be much in the common experience of fatigue which suggests that there may be something in the sensation of fatigue at the *central* level, not simply due to the fact that the muscles are generating sensations in themselves. I would summarize the sensation by the following analogy. It is as if the traffic is not moving. When curarized, I feel as though it's Sunday morning in the central nervous system! Whereas when I am fatigued, it's more like a Friday night traffic jam. I would add that the sensation is much clearer when I am totally curarized. If you are partially curarized you get sensations from the non-curarized muscles about the effort you are making. I imagine that Dr Merton has done some of these things to himself as well?

Merton: I have never been totally curarized, but I have been partially curarized twice. I agree that it is quite a different feeling from fatigue. With partial curarization you can feel that you are making a greater effort in order to move; this, you say, is not what happens when you are totally curarized.

Wilkie: Is the subjective sensation completely different from that of having a block higher up, such as in one's radial nerve? When I blocked my radial nerve in the spiral groove (Wilkie 1950) I was very surprised at the lack of any sensation of effort such as one encounters during real exercise.

Merton: I am not sure I could distinguish this feeling, which I also have experienced, from the feeling with partial curarization.

Campbell: The question is whether this difference in sensation arises at the level of the central organization of the motor discharge rather than, as most ordinary people would think, that the muscle is different when it has been working hard and when it is curarized.

Edwards: May we briefly pull together the relevant literature on curarization? This includes Moran Campbell's breath-holding experiments.

Campbell: If one holds one's breath as long as one can, particularly at resting lung volume, one develops an unpleasant sensation in the chest that can be relieved in various ways. This sensation is totally prevented by curarization. This observation suggests that the sensation of breath-holding requires a motor response for its development. It is not a sensation that arises as the result of the respiratory drives. It is a strange sensation, but as breathing is essentially a voluntary motor act (or the most involuntary of the various voluntary acts!) it may have its counterpart in other movements (Campbell et al 1969).

Edwards: Asmussen (1967) did studies of bicycle ergometry with partial curarization in which effort appeared subjectively to be greater. There was also a study by Molbech & Johansen (1969) on sustained submaximal isometric contractions in partially curarized subjects. The subjective comment was that the contractions had collapsed but there was no particular sensation of effort. I don't understand these authors' assertions that the sensations are different with decamethonium and curare. They suggested that the muscle spindles of the red and white muscle fibres making up the quadriceps muscle responded differently to the two drugs. In our attempt to relate effort to force of contractions under partial curarization, we were unable to detect an increase in effort. The effort paralleled the force; there was no evidence of a 'central effort factor' (Campbell et al 1977). That has not been supported by Gandevia & McCloskey (1977). Other studies support the idea that afferent information is important and that in its absence there is no central perception of effort (Roland & Ladegaard-Pedersen 1977).

There are thus conflicting reports about the roles of central and peripheral perception of effort in the overall sensation of fatigue in exercise. It would be interesting to know whether the perception of the effort is related to the control of the firing frequency in sustaining muscular activity, but before that can be understood we need information on the role of afferents from muscle.

Campbell: We have done an experiment in which untrained subjects sustained a near-maximal inspiratory pressure for as long as they could. They scaled the pressure they thought they were sustaining in a way which matched very well the pattern of the developed pressure. They also, with coaching (it's difficult to say how important the coaching was), distinguished this pressure from the effort they were making. The subjective effort increased but the subjective and objective force decreased (Campbell et al 1980).

Macklem: That is quite different from the experience that Leith & Bradley (1976) report in people breathing at maximal voluntary ventilation, isocapni-cally, where as their inspiratory muscles began to fatigue there was a progressive fall in ventilation. The subjects were unaware of the loss of pressure development and there was no sensation of dyspnoea.

Edwards: Could we remind ourselves of the electromyographic changes associated with failure of force, in conditions of partial curarization? I showed in my introductory paper (p 1-18) that there was high frequency fatigue in these circumstances. We know that there may be failing action potential; this has been used as a test by anaesthetists to monitor the degree of curarization during surgical operations (Heisterkamp et al 1969). Dr Grimby, have you looked at electromyographic changes in partial curarization?

Grimby: No. But when we did our blockade experiments in which only a few motor units remained we were surprised that we could drive our motor units at very high rates without the expected feeling of effort. Patients with partial nerve palsies also experience this. Patients with just a few motor units left after peripheral nerve lesions can drive these motor units at a high rate without any obvious effort, for some minutes. In normal circumstances such a high firing rate had required maximal effort. The difference in the sense of effort required to produce a high firing rate may be due to the fact that no tension was produced in the abnormal states discussed, whereas maximal tension developed in the normal state.

Edwards: Miller & Sherratt (1978) looked at the firing frequencies after partial denervation and found that firing frequencies decreased rapidly. Are you talking about a nerve trunk lesion or a peripheral neuropathy?

Grimby: A nerve trunk lesion. We presume that the central connections are relatively well preserved.

Wiles: We have mentioned that afferent information from muscle or its attachments may be important in regulating firing frequency. There may also be an important element of central control. In ballistic contractions, electro-myographic activity is thought to occur in the antagonist muscles before any force is produced in the agonists. This suggests that central modulation of the efferent outflow may potentially occur well in advance of any afferent modulation.

Edwards: The unmyelinated endings in muscle were mentioned. The other factor that comes into the discussion of fatigue and its perception may be a measure of discomfort in the muscle. We are working on muscle pain, and I am interested in the observations derived from single unit recordings from unmedullated fibres from Robert Schmidt's laboratory. An interesting relationship was found between metabolic and mechanical factors, in terms of the firing frequency over unmedullated (pain) fibres (Kniffki et al 1978). In prolonged ischaemia the firing frequency of non-medullated endings from muscle increases. If the muscle is made to contract the firing frequency increases considerably, so the amount of afferent traffic increases greatly for the degree of metabolic change (whether it is brought about by accumulation of potassium ions or peptides). Our recent studies on human muscle (Mills et al 1980) suggest that there is a sensing mechanism responsive both to metabolic factors and to tension.

Newsholme: I will try to relate what we discussed earlier—the metabolic changes—with what we are discussing now, the electrical activity. If we take the ATP/ADP phosphorylation–dephosphorylation system in muscle, this is a high capacity 'process' that provides energy for contraction and other processes. The ATP concentration is known to be very stable, and the changes in ADP concentration may be quite small (how small, we are still debating). If we want to use the change in the ATP/ADP concentration as a signal, we could argue that there should be something that can amplify this change in the ATP/ADP system to act as a metabolic signal. One system known to amplify the change is the formation of AMP through adenylate kinase, a near-to-equilibrium system.

We can go one stage further. Muscle contains the enzyme 5'-nucleotidase, which produces adenosine from AMP. The adenosine can then be converted to inosine. We have been studying this in heart muscle and all our evidence is that, if the rate of ATP turnover is 100, the leak to AMP and adenosine is only about 1%. However, it can detect a change in the concentration ratio of ATP/ADP and markedly increase the concentration of adenosine—a metabolite and physiological signal in the heart.

I am using adenosine purely to illustrate the principle of what might happen, but this is not totally removed from physiological reality. It is well established that adenosine is a vasodilator in the heart and that it works through this mechanism (Arch & Newsholme 1978). The adenosine produced in this way escapes from the heart muscle cell, attaches to a receptor on the smooth muscle of the arterioles, and causes the latter to vasodilate, allowing more blood to flow through the heart muscles, so providing more fuel and oxygen. This allows ATP to be regenerated at a greater rate. In heart, at least, there is evidence that adenosine can function as a link between metabolism and the physiology—the blood flow. Is it possible that adenosine, or something equivalent to it, could act at receptors in the muscle, to modify nervous information that is transmitted to higher centres?

Saltin: We have been thinking along the same lines. We haven't looked at adenosine. We have looked at inorganic phosphate, and osmolality, linked to increases in lactate and potassium concentrations. Our model is knee extension (50% of MVC). So far, potassium concentration is probably the factor most closely linked to subjective effort. That could also serve the same purpose as you describe in heart muscle. Potassium influences smooth muscle, and it influences the free nerve endings, and its effects could therefore be transmitted back centrally.

Edwards: Inorganic phosphate and ATP are released from contracting muscle. Forrester (1972) studied ATP as a factor for dilating blood vessels.

Saltin: What is puzzling is that if we make a very intense contraction, close to maximal, and hold it for 30 s or a minute, blood flow to the muscle is zero or nearly so. It is quite a different sensation from when you make an intense contraction with a cuff blocking the blood flow. The ischaemic pain is felt when a cuff is applied but not when you make a maximal voluntary contraction, when we think there is very limited flow.

Wilkie: In relation to Dr Newsholme's suggestion about an adenosine pathway, Meyer & Terjung (1979) have claimed that the degree to which adenosine is deaminated to inosine and ammonia is different in different muscle types, at least in the rat. In a fast twitch muscle (gastrocnemius) half the adenosine could be degraded to inosine and, more remarkably, rebuilt fairly rapidly; on the contrary, in a slow twitch muscle (soleus) this scarcely happened at all. I know that in frog sartorii, breakdown of adenosine scarcely happens unless you damage the muscle (see Dydynska & Wilkie 1966). The fibre composition of different muscles is different, and the biochemical changes at this level are evidently very different too.

Edwards: The condition of congenital adenosine deaminase deficiency is associated with muscle pains and a failure to produce ammonia during exercise (Fishbein et al 1978). This may be a single-enzyme defect worth looking at in man from this point of view.

Karlsson: McCloskey & Mitchell (1972) have shown that potassium in muscle activates the C afferent (unmyelinated) fibres and consequently increases blood pressure.

Saltin: Wildenthal et al (1968) showed that an increase in potassium or in osmolality caused an increase in blood pressure and in respiration. Mense (1977) recorded single nerve activity in unmyelinated fibres of the dorsal root. Very many compounds and factors affect some of the free nerve endings, so the picture is complex.

REFERENCES

Asmussen E 1967 Exercise and the regulation of ventilation. Circ Res 20 & 21 suppl I:132-145

Arch JRS, Newsholme EA 1978 The control of the metabolism and the hormonal role of adenosine. Essays Biochem 14:82-123

Buller NP, Garnett R, Stephens JA 1978 The use of skin stimulation to produce reversal of motor unit recruitment order during voluntary muscle contraction in man. J Physiol (Lond) 277:1P-2P

Buller NP, Garnett R, Stephens JA 1980 The reflex responses of single motor units in human hand muscles following muscle afferent stimulation. J Physiol (Lond) 303:337-349

Campbell EJM, Godfrey S, Clark TJH, Freedman S, Norman J 1969 The effect of muscular paralysis induced by tubocurarine on the duration and sensation of breath-holding during hypercapnia. Clin Sci (Oxf) 36:323-328

Campbell EJM, Edwards RHT, Hill DK, Jones DA, Sykes MK 1977 Perception of effort during partial curarization. J Physiol (Lond) 263:186P-187P

Campbell EJM, Gandevia SC, Killian KJ, Mahutte CK, Rigg JRA 1980 Changes in the perception of inspiratory resistive loads during partial curarization. J Physiol (Lond) 309:93-100

Clamann HP 1969 Statistical analysis of motor unit firing patterns in a human skeletal muscle. Biophys J 9:1233-1251

Clamann HP, Robinson AJ 1980 Time dependent effects of sural nerve stimulation on extensor motor pool organization. Neurosci Abstr 6:392

Datta AK, Stephens JA 1979 The effect of digital nerve stimulation on motor unit interspike intervals recorded during voluntary contraction of first dorsal interosseous muscle in man. J Physiol (Lond) 292:16P-17P

Desmedt JE, Godaux E 1977 Ballistic contractions in man: characteristic recruitment pattern of single motor units of the tibialis anterior muscle. J Physiol (Lond) 264:673-693

Dydynska M, Wilkie DR 1966 The chemical and energetic properties of muscles poisoned with fluorodinitrobenzene. J Physiol (Lond) 184:751-769

Fishbein WN, Armbrustmacher VW, Griffin JL 1978 Myoadenylate deaminase deficiency: a new disease of muscle. Science (Wash DC) 200:545-548

Forrester T 1972 An estimate of adenosine triphosphate release into the venous effluent from exercising human forearm muscle. J Physiol (Lond) 224:611-628

Freyshuss U, Knutsson E 1971 Discharge patterns in motor nerve fibres during voluntary effort in man. 83:278-279

Gandevia SC, McCloskey DI 1977 Changes in motor commands, as shown by changes in perceived heaviness, during partial curarization and peripheral anaesthesia in man. J Physiol (Lond) 272:673-689

Garnett RAF, Stephens JA 1978 Changes in the recruitment threshold of motor units in human first dorsal interosseous muscle produced by skin stimulation. J Physiol (Lond) 282:13P-14P

Garnett RAF, Stephens JA 1980 The reflex responses of single motor units in human first dorsal interosseous muscle following cutaneous afferent stimulation. J Physiol (Lond) 303:351-364

Garnett RAF, Stephens JA 1981 Changes in the recruitment threshold of motor units produced by cutaneous stimulation in man. J Physiol (Lond) 311:463-473

Grimby L, Asmussen E 1979 Muscle fatigue. Med Sci Sports 11:313-321

Grimby L, Hannerz J 1976 Disturbances in the voluntary recruitment order of low and high frequency motor units on blockade of the proprioceptive afferent activity. Acta Physiol Scand 96:207-216

Hannerz J, Grimby L 1979 The afferent influence on the voluntary firing range of individual motor units in man. Muscle Nerve 2:414-422

Heisterkamp DV, Skovsted P, Cohen PJ 1969 The effects of small incremental doses of d-tubocurarine on neuromuscular transmission in anesthetized man. Anesthesiology 30:500-505

Kniffki K-D, Mense S, Schmidt RF 1978 Responses of group IV afferent units from skeletal muscle to stretch, contraction and chemical stimulation. Exp Brain Res 31:511-522

Leith DE, Bradley M 1976 Ventilatory muscle strength and endurance training. J Appl Physiol 41:508-516

McCloskey DI, Mitchell JH 1972 Reflex cardiovascular and respiratory responses originating in exercising muscle. J Physiol (Lond) 224:173-183

Marsden CD, Meadows JC, Merton PA 1971 Isolated single motor units in human muscle and their rate of discharge during maximal voluntary effort. J Physiol (Lond) 217:12P-13P

Mense S 1977 Nervous outflow from skeletal muscle following chemical noxious stimulation. J Physiol (Lond) 267:75-88

Meyer RA, Terjung RL 1979 Differences in ammonia and adenylate metabolism in contracting fast and slow muscle. Am J Physiol Cell Physiol 237:C111-C118

Miller RG, Sherratt M 1978 Firing rates of human motor units in partially denervated muscle. Neurology 28:1241-1248

Mills KR, Newham D, Edwards RHT 1980 Force, metabolism and contraction frequency as interactive determinants of ischaemic muscle pain. Clin Sci (Oxf) 59:13P

Molbech S, Johansen SH 1969 Endurance time in static work during partial curarization. J Appl Physiol 27:44-48

Roland PE, Ladegaard-Pedersen H 1977 A qualitative analysis of sensations of tension and of kinaesthesia in men. Evidence for a peripherally originating muscular sense of effort. Brain 100:671-692

Stalberg E, Trontelj JV 1979 Single fibre electromyography. Miravelle Press, Old Woking, Surrey, UK

Stephens JA, Garnett R, Buller NP 1978 Reversal of recruitment order of single motor units produced by cutaneous stimulation during voluntary muscle contraction in man. Nature (Lond) 272:362-364

Wildenthal K, Mierzwiak D, Skinner N Jr, Mitchell J 1968 Potassium induced vascular reflexes from the dog hindlimb. Am J Physiol 215:543-546

Wilkie DR 1950 The relation between force and velocity in human muscle. J Physiol (Lond) 110:249-280

Muscle fatigue due to changes beyond the neuromuscular junction

D. A. JONES

Department of Human Metabolism, Rayne Institute, University College London Medical School, University Street, London WC1 6JJ, UK

Abstract A number of changes in function occur beyond the neuromuscular junction during activity; three main types are described. (a) During high frequency stimulation there is a rapid loss of force accompanied by a slowing of the action potential waveform and an increase in the excitation threshold of the muscle. It is suggested that accumulation of K^+ in the extracellular spaces of the muscle may be responsible for these changes. (b) Slowing of relaxation is a feature of fatigued muscle. The slowing allows a reduction in activation frequency (minimizing high frequency fatigue) without resulting in an appreciable loss of force. (c) Changes in shape and amplitude of the twitch have considerable effects on the force generated by low frequencies of stimulation. After a brief tetanus there is a reduction in the width of the twitch which increases the fusion frequency of the muscle and may account for the 'sag' seen at the start of low frequency contractions. After a prolonged series of contractions the twitch amplitude is reduced and remains so for several hours. This may be the result of some structural damage to the sarcoplasmic reticulum or transverse tubular system.

Used colloquially the word 'fatigue' may mean many things, but even when restricted to muscle the single word may oversimplify the variety of changes in function that occur during prolonged or repeated contraction. Not all the changes are necessarily detrimental; some certainly lead to a loss of force but others may help to maintain it by counteracting the effects of the more harmful changes.

Alterations in function and activity occur at all levels in the chain of command controlling muscle contraction, but this paper is concerned only with those arising, for better or worse, in the muscle itself—that is, beyond the neuromuscular junction.

Three main types of change will be described: the response to high frequency stimulation, the slowing of relaxation, and changes in the ampli-

1981 Human muscle fatigue: physiological mechanisms. Pitman Medical, London (Ciba Foundation symposium 82) p 178-196

tude and shape of the twitch. The significance of these various changes will be discussed with particular reference to the way they may combine to influence the time course of a prolonged contraction.

Experimental methods

Methods used to record force and electrical activity of human muscle *in situ* were as described by Bigland-Ritchie et al (1979). Isolated mouse and rat muscles were studied as described by Jones et al (1979) except that the temperature was maintained at 35 °C. Massive stimulation was via platinum electrodes, 1 cm apart on either side of the muscle, square wave pulses being used, usually of 0.1 ms duration and 40 V. With this duration, 40 V was approximately three times the voltage required to generate maximum force. Samples of human muscle obtained during the normal course of surgery were prepared as described by Moulds et al (1977), incubated at 35 °C and stimulated in the same way as the rat and mouse muscles. Recordings of muscle action potentials and force were made from strips of rat diaphragm using an incubation chamber and an arrangement of stimulating and recording electrodes very similar to that used by Niedergerke (1956) to record from strips of heart muscle. The medium was maintained at 35 °C by circulating it through a small reservoir where it was warmed and oxygenated. In all experiments on isolated preparations the incubation medium contained tubocurarine at 2.0 mg/100 ml.

Changes during activity

Fatigue at high frequencies

When the human adductor pollicis muscle is stimulated via the nerve the time course of the subsequent loss of force depends very much on the frequency of stimulation. At high frequencies (around 80/s) force is maintained for only a few seconds before falling by about 30 s to between 10 and 20% of the initial value. In contrast, when the muscle is stimulated at 20/s the force remains nearly constant for 60 s or longer (Jones et al 1979). During prolonged contractions a greater force × time (the area under the force curve) can be obtained by stimulating at the lower frequencies. This specific loss of force at high frequencies will be referred to as 'high frequency' fatigue.

A substantial part of this force loss may be attributed to a failure of transmission across the neuromuscular junction. However, it is notable that isolated and curarized muscle preparations in which the muscle is directly stimulated show a similar sensitivity to high frequency stimulation (Jones et al 1979). In Fig. 1A is shown the result of stimulating the human adductor

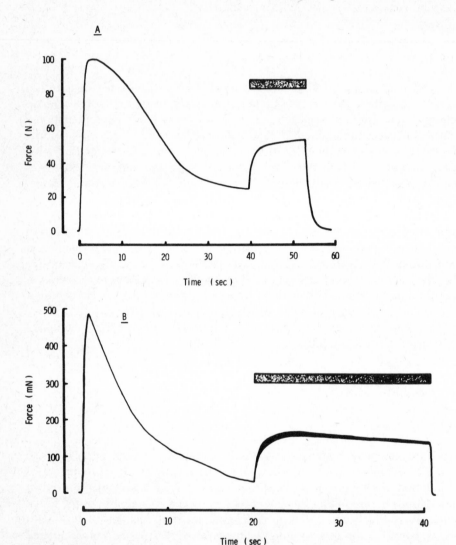

FIG. 1. High frequency fatigue of human muscle preparations *in situ* and *in vitro*.

A. Force generated by the adductor pollicis stimulated via the ulnar nerve at 100/s for 40 s when the frequency was reduced to 20/s.

B. Force generated by an isolated and curarized preparation of human quadriceps muscle directly stimulated at 100/s for 20 s when the frequency was reduced to 20/s.

In both illustrations the shaded bars indicate stimulation at 20/s.

pollicis via the nerve at 100/s and then reducing the frequency to 20/s. Changing the frequency resulted in a rapid increase in force which was well maintained. An isolated preparation of human quadriceps muscle was similarly stimulated at 100/s and showed a rapid loss of force which was reversed when the stimulation frequency was reduced to 20/s (Fig. 1B). This demonstrates that high frequency fatigue can be the result of postsynaptic changes.

The very rapid recovery of force in the isolated muscles makes it unlikely that it is associated with resynthesis of phosphoryl creatine, as this has a half-time of around 30 s (Harris et al 1976). Instead it suggests some rapid process, such as diffusion, and it has been proposed that changes in the inter-fibre cation concentrations (increased K^+ and/or decreased Na^+) may occur during high frequency stimulation and be responsible for the loss of force (Bigland-Ritchie et al 1979). Such changes would be rapidly reversed during a pause in stimulation or a reduction in frequency, as the cation concentrations return to normal by diffusion into and out of the surrounding fluids.

Adrian & Peachy (1973) estimated that for every muscle action potential the Na^+ concentration in the T (transverse) tubes may fall by 0.5 mM and the K^+ concentration may increase by 0.28 mM. Since the normal external K^+ concentration is relatively low, the efflux of K^+ may very rapidly lead to a significant increase. Experimental reduction of the Na^+ concentration in the medium promotes the development of high frequency fatigue in isolated mouse muscle preparations (Jones et al 1979) and preliminary experiments have shown that increasing the concentration of K^+ in the medium has a similar effect.

A characteristic of high frequency fatigue of human muscle stimulated *in situ* is a slowing of the conduction velocity of the muscle action potential, so that the surface recorded action potential becomes prolonged, loses amplitude and eventually fails (see Fig. 2A and Bigland-Ritchie et al 1979). Possible causes of this were examined by recording action potentials from strips of rat diaphragm, curarized and directly stimulated at high frequencies. During stimulation at 50/s, changes were seen in the waveform of the action potential recorded from the diaphragm strip (Fig. 2B) that were very similar to those noted during stimulation of human muscle *in situ* (Fig. 2A).

In a fresh preparation action potentials were recorded at intervals in response to single shocks and then sufficient KCl was added to the medium reservoir to bring the K^+ concentration from 5 mM to 10 mM when the medium in the two chambers was fully mixed. Thus for a period of about 5 min the K^+ concentration around the muscle was steadily increasing. During this time action potentials evoked by single shocks were recorded. Increasing the external K^+ was found to decrease the amplitude and slow the time course

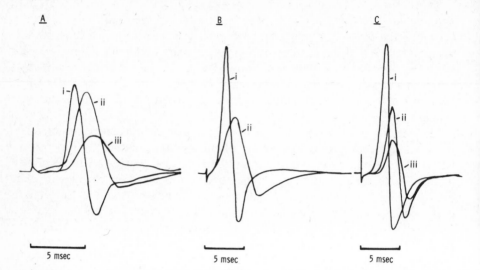

FIG. 2. Slowing of the action potential waveform of muscle preparations *in situ* and *in vitro*.
A. Action potentials recorded from adductor pollicis after (i) 2 s, (ii) 20 s and (iii) 40 s of stimulation of the ulnar nerve at 50/s.
B. Action potentials of an isolated and curarized strip of rat diaphragm after (i) 1 s and (ii) 10 s of stimulation at 50/s.
C. Action potentials from an isolated rat diaphragm preparation in response to single shocks at intervals while the K^+ concentration of the medium was increased from (i) 5 mM to (iii) 10 mM.

of the recorded action potential (Fig. 2C). This was reversed when the normal K^+ concentration in the medium was restored. These changes are very similar to those seen during high frequency fatigue of muscle both *in situ* and *in vitro* (Fig. 2A, B).

Another change during high frequency fatigue of isolated muscles is an alteration in the excitation threshold. With the arrangement of electrodes used and 40 V, maximum twitch tension could be obtained with pulses of 0.02 ms duration and the force remained constant for pulse durations up to 0.2 ms. Tetrodotoxin reduced this force to less than 1% of the initial value with pulses of this duration, indicating that the contractions were produced by a normal mechanism involving action potentials. With rat soleus muscle brief tetani at 200/s with pulses of 0.2 and 0.02 ms gave contractions within 10% of one another in fresh muscle (Fig. 3A). However, after 20 s stimulation with the shorter pulses this was no longer the case, since increasing the pulse duration resulted in a recovery of force. Similar experiments with diaphragm preparations have shown that this was accompanied by (and is assumed to be the result of) an increase in the action potential amplitude. The findings were

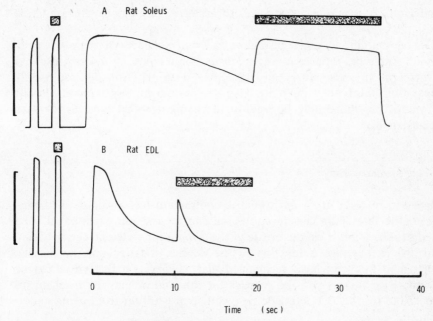

FIG. 3. Excitation threshold during high frequency stimulation.
A. Isolated and curarized rat soleus muscle.
B. Isolated and curarized rat extensor digitorum longus muscle (EDL).
Muscles were stimulated at 200/s with pulse of 40 V and 0.02 ms duration. During the periods
indicated by the shaded bars the pulse duration was increased to 0.2 ms. In the fresh muscle this
made little difference but after a period of stimulation there was a marked increase in force. Bars
to the left of the records indicate forces of 1 N.

not due to polarization of the electrodes, as reversing the polarity did not
affect the force. Isolated human muscle preparations showed a similar pattern
to the rat muscle.

Change in excitation threshold as the result of high frequency stimulation
has been demonstrated by Krnjević & Miledi (1958), who suggested that it
might be associated with changes in the extracellular cation concentrations.
This would fit with the effects of experimentally altering the external cation
concentrations on the rate of force fatigue and on the form of the muscle
action potential.

Accumulation of K^+ in the T tubes and inter-fibre spaces could increase the
excitation threshold of the muscle membrane, slowing the conduction of the
action potential which may eventually fail to propagate along the surface of
the muscle and thus result in a loss of force. Although this may seem a likely
scenario, it has yet to be shown that the inter-fibre K^+ concentration does rise

significantly and that increased K+ does cause a change in the excitation threshold.

The fast rat muscle, extensor digitorum longus, behaves in a similar way to the slow soleus but fatigues more rapidly and shows only a transitory response to increases in pulse duration (Fig. 3B). With the isolated fast muscle preparations this type of stimulation leads to permanent loss of force, possibly indicating that there may be additional changes in fast muscles in these circumstances.

Slowing of relaxation

Relaxation of force from an isometric contraction has a well-defined time course, the later part closely approximating to a single exponential. The speed of relaxation is characteristic of the 'type' of muscle and the species it came from; it is also a function of the degree of fatigue of the muscle. Although it has long been recognized that slowing is a feature of fatigue (Mosso 1915, Feng 1931) the reasons for this and indeed the mechanisms responsible for the characteristic time course are still far from being understood.

The half-time of the exponential phase typically increases 2–3-fold as the result of fatiguing voluntary contractions (Edwards et al 1972) and similar changes occur as the result of stimulated contractions of isolated animal muscles (Edwards et al 1975a). Under anaerobic conditions there is little (isolated animal muscle) or no (human muscle *in situ*) recovery of the slowing. When circulation is restored to the human muscle the recovery has a half-time of about 30 s (Edwards et al 1972, Wiles 1980) which resembles the time course of phosphoryl creatine resynthesis (Harris et al 1977) rather than of the removal of lactate from the muscle (Diamant et al 1968, Hermansen & Osnes 1972).

As for the mechanism determining the time course of relaxation and, presumably, the mechanism affected in fatigue, two main possibilities have been considered. These are: (a) that the relaxation reflects the rate of removal of calcium from the interior of the fibre and (b) that it reflects the time course of dissociation of cross-bridges after the activating calcium has been removed.

Calcium pumping. It is widely held that the kinetics of the accumulation of calcium by the sarcoplasmic reticulum determines the rate of relaxation (e.g. Sandow 1965), but direct evidence for this is slight. Studies of intracellular free calcium using the luminescent protein aequorin show the light signal to be a very rapid transient. During a twitch of a single frog muscle fibre the light signal returned to near baseline by the time the twitch force had reached its peak (Blinks et al 1978). The relationship between light signal, free calcium, calcium bound to troponin and the development of force is inevitably very complex.

Changes in cross-bridge cycling. Relaxation of force from an isometric contraction involves the dissociation of myosin cross-bridges, a process that requires the binding of ATP to the myosin molecule. It is possible that dissociation of the cross-bridges may be the limiting process in relaxation which is slowed in fatigued muscle. This possibility was examined by Edwards et al (1975a). Although comparisons of calculated cross-bridge turnover rates and relaxation proved unsatisfactory, one test of the hypothesis emerged which still seems valid. A decreased rate of cross-bridge dissociation should be reflected in a reduced rate of turnover of the cross-bridges, and therefore of ATP, during the contraction; so the slower the relaxation, the slower the muscle ATP turnover.

In experiments with mouse soleus muscles such a change was seen; the turnover calculated from changes in ATP, phosphoryl creatine and accumulation of lactate fell to about half at times when the half-time of relaxation was prolonged 2–3-fold. Considering the relatively small changes in ATP concentration in fatigued muscle and the high affinity of the actomyosin ATPase for ATP (Yagi & Mase 1965) it seemed unlikely that the slow relaxation was caused by a reduced amount of ATP as a substrate for the actomyosin ATPase but rather that it might be responsible for changes in possible regulatory subunits, such as the myosin light chains (Edwards et al 1975a).

Recently, Dawson et al (1980) have used nuclear magnetic resonance (n.m.r.) techniques to follow metabolite changes in fatiguing frog muscle. They found no change in the rate of ATP turnover per unit force at times when the relaxation was slowed, indicating that in this preparation cross-bridge turnover was not affected. Similar n.m.r. studies of isolated mammalian muscles (rat or mouse) might be a suitable way of resolving this apparent conflict.

Evidence to support the idea of a reduced turnover of cross-bridges comes from studies of heat production by muscle during prolonged contractions in man. Heat production was measured with an intramuscular probe during both voluntary and stimulated contractions. It was found that the heat production rate, expressed per unit force, fell towards the end of the contraction when relaxation was slowed. Depending upon the proportions of heat derived from phosphoryl creatine splitting and glycolysis, it was calculated that the reduced heat production rate might correspond to a 2–4-fold reduction in ATP turnover (Edwards et al 1975b, Edwards & Hill 1975, Wiles 1980).

In normal muscles the slowing of relaxation is accompanied by reductions in ATP and phosphoryl creatine as well as the accumulation of lactate and H^+ (Edwards et al 1975a, Dawson et al 1980). However, experiments with isolated muscles poisoned with iodoacetate (Edwards et al 1975a) and our experience of patients with myophosphorylase deficiency (Wiles et al 1981), both of which conditions can show a slowing of relaxation, indicate that the formation and accumulation of lactate is not the immediate cause of the slowing.

Free ADP will increase during a contraction, but preliminary experiments have shown no association between this and the rate of relaxation. The most likely 'causative agent' thus appears to be a reduced ATP level, but the possibility of some allosteric effect of either free creatine or phosphoryl creatine (both of which undergo large changes during contractions) should not be forgotten.

The decrease in ATP concentration may be directly responsible for a reduced rate of calcium pumping by the sarcoplasmic reticulum; alternatively it may produce changes in the phosphorylated subunits of myosin or troponin, influencing either the kinetics of the cross-bridge or the affinity of calcium binding by troponin.

Changes in twitch force after activity

The size and shape of the muscle twitch is very variable. The best-known change is post-tetanic potentiation (Close & Hoh 1968, Hanson 1974) where the twitch force is found to be increased after a brief tetanic stimulation. Study of the phenomenon is still at the descriptive and speculative stage and it is likely that more than one type of change occurs.

Isolated mouse fast and slow muscles respond differently to brief tetanic stimulation. The fast extensor digitorum longus (EDL) shows potentiation of twitch force after a 1s tetanus (Fig. 4A) and a marked speeding of the relaxation so that the width of the twitch is reduced. With the slow soleus muscle, brief tetani caused only depression of the peak force but the same change in width occurred as seen with the EDL (Fig. 4B).

Human muscle stimulated *in situ* with brief tetani shows a potentiation of force and a similar change in shape of the twitch (Fig. 4C).

After longer tetani with isolated muscles of both types only a depression of twitch force was seen. This contrasts with the human muscle *in situ* where longer stimulation resulted in potentiation of the twitch and a slower time course. A similar finding is reported by Close & Hoh (1968) for *in situ* animal muscle, suggesting that there may be a difference, in this respect, between the behaviour of isolated and *in situ* preparations.

A different type of change in the twitch is seen during oxidative recovery

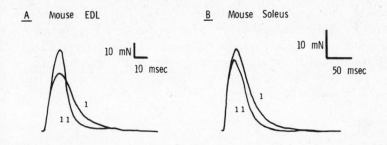

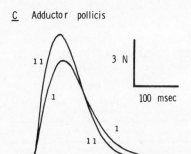

FIG. 4. Twitch amplitude and shape after brief tetanic contractions.
A. Mouse extensor digitorum longus (i) before and (ii) 1 s after a 1 s tetanus at 100/s.
B. Mouse soleus muscle (i) before and (ii) 1 s after a 1 s tetanus at 100/s.
C. Human adductor pollicis stimulated via the ulnar nerve (i) before and (ii) 1 s after a 2 s tetanus at 100/s.

from mildly fatiguing contractions. During about the first 5 min the twitch becomes larger with a very long relaxation phase. The prolonged twitch returns to normal over 10–30 min but this is speeded by brief tetani or by making the muscle anaerobic. This phenomenon has been described in both isolated mouse soleus muscle and human adductor pollicis stimulated *in situ* (Hill & Jones 1978).

The various changes in the twitch described above are restored to normal in the course of minutes or tens of minutes, indicating some mechanism linked to changes in fibre metabolites or cation concentrations. After severe exercise, however, a different type of change in the twitch occurs which is much longer lasting and may be caused by some structural change in the fibre.

After a long series of ischaemic contractions of the adductor pollicis it was found that while the maximum tetanic force was restored to normal by about 30 min, the force generated at lower frequencies took many hours to return (Edwards et al 1977). This type of fatigue will be referred to as 'low frequency' fatigue. The same type of behaviour is seen with isolated animal muscles. Both fast and slow muscles show the effect although it requires more stimulation to produce the same degree of fatigue in a soleus as in an EDL.

After repetitive stimulation of a mouse soleus muscle the high frequency force returned rapidly to 85% of the initial value but the force at 30/s had recovered to only 58% after 60 min (Fig. 5). Underlying, and responsible for,

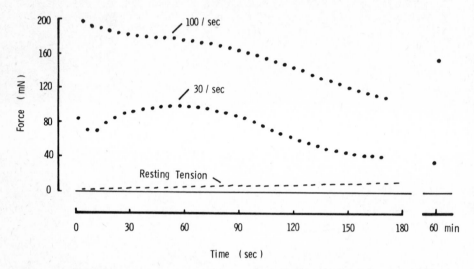

FIG. 5. Forces generated by different frequencies of stimulation during a series of contractions and after recovery. Mouse soleus muscle was stimulated every 5 s with a 1 s tetanus at 30/s followed, after a 1 s pause, by a 1 s tetanus at 100/s. After nearly 3 min the muscle was allowed to recover for 60 min and the force generated at the two frequencies was measured again. Resting tension between the tetani is indicated. The solid line is the resting tension of the fresh muscle.

the reduced force at 30/s was a small twitch. After 60 min recovery the twitch had returned to 60% of the initial peak force (Fig. 6).

During low frequency fatigue the relatively normal force at high frequencies suggests that the contractile proteins retain their ability to generate force—and it was found that the muscle action potential remains normal. Biopsy studies have shown the muscle metabolites to be restored when the low frequency force was still reduced (Edwards et al 1977). These findings suggest that the defect may be either in the quantity of calcium released from the sacroplasmic reticulum in response to a single action potential or a change in the affinity of the troponin binding site for calcium. Both defects would have the effect of reducing the twitch force while at higher stimulation frequencies, when the interior of the fibre is saturated with calcium, producing relatively normal force.

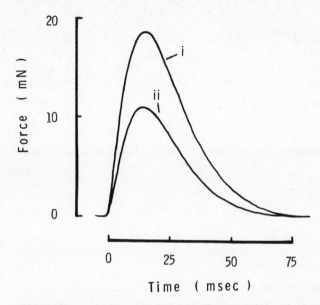

FIG. 6. Change in the twitch during low frequency fatigue in the mouse soleus muscle. Twitches were recorded during the experiment illustrated in Fig. 5. (i) Twitch of the fresh muscle, (ii) recorded after 60 min recovery at a time when the tetanic force at 30/s was reduced compared to the force at 100/s.

The very slow recovery suggests possible structural damage to the T system or sarcoplasmic reticulum membranes rather than changes in the calcium binding by troponin, but no evidence of damage has been seen. Electron microscope studies of fatigued muscle have shown a few cases of swelling of the T tubes but the most common change is a swelling and disruption of the mitochondria. The significance of these changes is not clear and none has been correlated with the extent of fatigue.

The effects of activity on the twitch may thus be categorized according to the severity of the preceding contractions. After brief tetani there is potentiation of twitch force in fast muscles but depression in slow. Both muscle types show a shortened twitch and changes return to normal in a matter of minutes. After contractions of intermediate duration the twitch may become prolonged during oxidative recovery, returning to normal over tens of minutes. After heavy exercise the twitch is reduced in size and recovers only very slowly over a period of hours.

The major influences on the force maintained during a prolonged contraction may be summarized and illustrated by considering the course of a single tetanus (Fig. 7). Rat diaphragm is a fast muscle, so that at a stimulation frequency of 50/s about 60% of the maximum force is generated and, in the

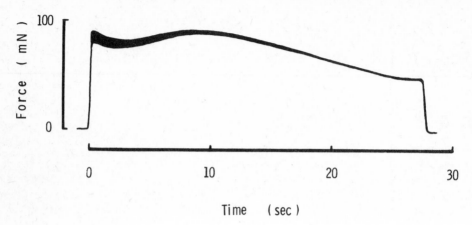

FIG. 7. Force recorded from a rat diaphragm preparation stimulated at 50/s. The oscillation of the force record is indicated by the width of the darker top of the record. This became less during the course of the contraction as the muscle relaxation slowed.

fresh muscle, there is an appreciable oscillation on the force record. In the first two seconds the force falls and if the contraction were stopped and the shape of a single twitch examined it would be found to be reduced in width. The reduction in tetanic force at this time is due to an increased oscillation and fusion frequency. In the next 10 s the oscillation decreases and the mean force rises as the muscle slows and the fusion frequency decreases. Thereafter the force decreases, probably for a number of reasons. In this contraction, stimulating at 50/s, a degree of high frequency fatigue will have developed. Evidence for this is the change in shape of the action potential, that illustrated in Fig. 2B (p 182) being recorded at 10 s from this preparation. Changes are also occurring which, if the muscle were allowed to recover, would be evident as low frequency fatigue. Thus during the later part of the contraction not only is the action potential not propagating to all parts of the muscle but the quantity of calcium released in response to it is also decreased. This latter change is to some degree minimized by the slower relaxation of the muscle.

Acknowledgements

As the references cited will show, much of the work discussed has been done in close collaboration with Professor R. H. T. Edwards, Professor D. K. Hill and Dr B. Bigland-Ritchie, to whom I am most grateful. The work has been generously supported by the Muscular Dystrophy Group of Great Britain and by The Wellcome Trust.

REFERENCES

Adrian RH, Peachy LD 1973 Reconstruction of the action potential of frog sartorius muscle. J Physiol (Lond) 235:103-131

Bigland-Ritchie B, Jones DA, Woods JJ 1979 Excitation frequency and muscle fatigue: electrical responses during human voluntary and stimulated contractions. Exp Neurol 64:414-427

Blinks JR, Rüdel R, Taylor SR 1978 Calcium transients in isolated amphibian skeletal muscle fibres: detection with aequorin. J Physiol (Lond) 277:291-323

Close R, Hoh JFY 1968 The after effects of repetitive stimulation on the isometric twitch contraction of rat fast skeletal muscle. J Physiol (Lond) 197:461-477

Dawson MJ, Gadian DG, Wilkie DR 1980 Mechanical relaxation rate and metabolism studied in fatiguing muscle by phosphorus nuclear magnetic resonance. J Physiol (Lond) 299:465-484

Diamant B, Karlsson J, Saltin B 1968 Muscle tissue lactate after maximal exercise in man. Acta Physiol Scand 72:383-384

Edwards RHT, Hill DK 1975 'Economy' of force maintenance during electrically stimulated contractions of human muscle. J Physiol (Lond) 250:13-14P

Edwards RHT, Hill DK, Jones DA 1972 Effect of fatigue on the time course of relaxation from isometric contractions of skeletal muscle in man. J Physiol (Lond) 227:26-27P

Edwards RHT, Hill DK, Jones DA 1975a Metabolic changes associated with the slowing of relaxation in fatigued mouse muscle. J Physiol (Lond) 251:287-301

Edwards RHT, Hill DK, Jones DA 1975b Heat production and chemical changes during isometric contractions of the human quadriceps muscle. J Physiol (Lond) 251:303-315

Edwards RHT, Hill DK, Jones DA, Merton PA 1977 Fatigue of long duration in human skeletal muscle after exercise. J Physiol (Lond) 171:769-778

Feng TP 1931 The heat-tension ratio in prolonged tetanic contractions. Proc R Soc Lond B Biol Sci 108:522-537

Hanson J 1974 The effects of repetitive stimulation on the action potential and twitch of rat muscle. Acta Physiol Scand 90:387-400

Harris RC, Edwards RHT, Hultman E, Nordesjö L-O, Nylind B, Sahlin K 1976 The time course of phosphorylcreatine resynthesis during recovery of the quadriceps muscle in man. Pflügers Arch Eur J Physiol 367:137-142

Hermansen L, Osnes JB 1972 Blood and muscle pH after maximal exercise in man. J Appl Physiol 32:304-308

Hill DK, Jones DA 1978 Prolongation of mammalian muscle twitches during oxidative recovery from contraction. J Physiol (Lond) 280:66P

Jones DA, Bigland-Ritchie B, Edwards RHT 1979 Excitation frequency and muscle fatigue: mechanical responses during voluntary and stimulated contractions. Exp Neurol 64:401-413

Krnjević K, Miledi R 1958 Failure of neuromuscular propagation in rats. J Physiol (Lond) 140:440-461

Mosso A 1915 Fatigue, 3rd edn. Drummond M, Drummond WG (transl) Allen & Unwin, London, p. 78-80

Moulds RFW, Young A, Jones DA, Edwards RHT 1977 A study of the contractility, biochemistry and morphology of an isolated preparation of human skeletal muscle. Clin Sci Mol Med 52:291-297

Niedergerke R 1956 The 'staircase' phenomenon and the action of calcium on the heart. J Physiol (Lond) 134:569-583

Sandow A 1965 Excitation–contraction coupling in skeletal muscle. Pharmacol Rev 17:265-320

Wiles CM 1980 The determinants of relaxation rate of human muscle *in vivo*. PhD thesis, University of London

Wiles CM, Jones DA, Edwards RHT 1981 Fatigue in human metabolic myopathy. This volume
    p 264-277
Yagi K, Mase R 1965 Possible compartmentation of adenine nucleotides in coupled reaction
    system composed of F-actomyosin-adenosine-triphosphatase and creatine kinase. In: Ebashi S
    et al (eds) Molecular biology of muscular contraction. Elsevier, Amsterdam, p 109-123

DISCUSSION

Merton: The experiments in which you stimulated curarized muscle directly were very interesting, but I wonder whether the explanation in terms of a building-up of the extracellular potassium concentration or a lowering of the sodium concentration is the most likely one. Lüttgau (1965) studied fatigue of action potentials in single frog fibres in a bath of Ringer. He concluded that 'the rapid decline of the action potential during a tetanus . . . cannot be due to accumulation of sodium or to loss of potassium, since the changes in concentration resulting from the ionic movements during activity would only begin to have a perceptible effect after hundreds of impulses'. Such action potential fatigue was not seen in non-contractile fibres poisoned with cyanide or iodoacetate. This remarkable observation suggested that 'the fatigue of the action potential is caused by reactions connected with contraction and its activation'.

Jones: I am sure there are additional ways in which the muscle becomes inexcitable. One interesting possibility might be some type of change similar to that responsible for the development of inexcitability in hyper- or hypokalaemic periodic paralysis. As can be seen in Fig. 3 (p 183), fast and slow muscle have very different fatigue characteristics, and I think it very likely that there may be additional mechanisms contributing to the loss of force in the fast muscles.

Dawson: In later studies (Grabowski et al 1972, Oetlicker 1973, Fink & Lüttgau 1976) Lüttgau and his co-workers specifically measured changes in the membrane characteristics of fatigued muscles. They found a marked decrease in membrane resistance and an increase in potassium conductance. However, their measurements were made in severely exhausted fibres and thus do not constitute evidence that the decline in force results from changes in membrane characteristics in studies similar to those of Dr Jones.

Jones: I have tried Lüttgau's (1965) experiment with poisoned rat diaphragm and could not repeat his finding (unpublished observations). When the muscle produced no more force, there were no more action potentials, either.

Dawson: I am very interested in how the decline of force in high frequency fatigue might be related to the changes in the action potential. In your low sodium experiments you showed changes in force but you didn't show any changes in the action potential. Conversely, in the high potassium experiments, you showed changes in action potential but not the force. So in each case there is a missing link.

Jones: I have not, so far, measured the two at the same time. But, comparing the time course of the separate experiments, the action potential does start to fail at about the same time as the force. It seems likely that the increase in potassium results in a failure of the action potential and subsequent loss of force.

Saltin: In humans doing 50% maximal voluntary contractions of the knee extensor muscles, potassium builds up in the interstitial space of the muscle. The subjects were told to maintain 50% MVCs for 1–1½ minutes. Flow was measured in the femoral vein, and potassium concentration. During the contraction there is a rise from 4.1 to 6.5 m equiv/l potassium. When the contraction is released flow immediately increases to 3 or 4 l/min. During the next 5–10 seconds there is a marked, transient loss of potassium from that area. Values are normal again within 10 seconds. This rapidity of recovery is evidence that the potassium level was increased in the interstitial space during the contraction.

Wiles: There is no evidence that 'high frequency fatigue' actually occurs during voluntary contractions in normal human muscle, so these raised extracellular potassium concentrations must be tolerated by the sarcolemma.

Jones: I would like to stress that I do not think 'high frequency' fatigue, as I have described it, ever occurs during normal voluntary contractions. Our work with Dr Brenda Ritchie (Jones et al 1979, Bigland-Ritchie et al 1979) has shown that during voluntary contractions the natural firing frequency rapidly declines. It appears that there is a mechanism, one function of which may be to avoid 'high frequency' fatigue. Nevertheless, Professor Saltin's remarks are very interesting because they indicate that there are large fluxes of K^+ in the contracting muscle, so that if the muscle were artificially stimulated at high frequency, accumulation of K^+ could well become an important cause of failure.

Edwards: You did not distinguish in your paper between the effects of altering ion concentrations in the interstitial space and in the transverse (T) tubular system. I think we should make this distinction. The changes that will alter conduction velocity, and the shape and size of the EMG, will be those dominated by alterations in the surface membrane of the fibre and they would be influenced by the ionic concentrations in the interstitial space. The effects of interfering with the conduction of the action potential down the tubular system would be different. You could still have a membrane action potential but the main effect would be impaired excitation–contraction coupling. Bezanilla et al (1972) did electron micrographic studies in which, when the external sodium concentration was changed, they found not the nice register of contraction about the Z lines, showing uniform contraction across the muscle fibre, but buckling of the internal myofibrils, indicating that these had not been fully activated. As we explore and clarify the practical separation between high and low frequency fatigue, it may be worth distinguishing whether the problem is primarily in the interstitial space or in the tubular system.

Jones: I would not make such a clear distinction between changes in the lumen of the T tubes and in the interstitial space. So far as high frequency stimulation is concerned, I imagine K^+ accumulating first, and to the greatest extent, in the depths of the T tube system, leading to a failure of propagation. With continued stimulation this would spread to involve the interstitial spaces and the surface membranes of the fibres.

Edwards: Can you say that? If you postulate this, then low frequency fatigue should always come before high frequency fatigue. That is, impaired excitation–contraction coupling, or impaired contractility, should show itself before failure of the surface-recorded action potential.

Jones: I am not at all sure that failure of conduction down the T tubes will result in low frequency fatigue. It would be interesting to see whether a muscle in a medium containing a high K^+ or low Na^+ concentration showed any evidence of low frequency fatigue, as well as an increased tendency to lose force at high frequency.

Edwards: When Bezanilla et al (1972) showed impaired propagation there were striking changes from one part of the same cell to the other.

Jones: But we do not know whether the preparation showed high or low frequency fatigue. As I said before, it would be very interesting to investigate this.

Although it is unlikely that high frequency fatigue occurs during normal voluntary contractions it does become important during studies of fatigue when we are attempting to measure maximum force generation with test stimulation. To obtain maximum force in the fresh mammalian muscle working at body temperature it is necessary to use relatively high frequencies. However, during the course of a fatiguing contraction there comes a point where the high frequency test stimulation itself results in a rapid loss of force. At this time the maximum (but now lower) force can only be generated by stimulating at a lower frequency (Jones et al 1979). In these circumstances it is no longer clear whether the loss of maximum force is the result of reduced contractility or a consequence of the lower stimulation frequency that has to be used. What is needed to answer this question is some way of activating the muscle that does not involve generating action potentials in the surface membranes. With isolated preparations, potassium or caffeine contractures are a possibility but are probably of too slow a time course to use with whole muscles. An alternative might be to stimulate with pulses of either very high voltage or long duration which can activate the muscle by a mechanism that is not sensitive to tetrodotoxin (personal observations).

Edwards: This problem came out clearly in our central fatigue experiments (Bigland-Ritchie et al 1978). It is worth emphasizing that from the fact that force returns when you reduce the frequency, and from the time course of this recovery, the loss of force cannot be simply attributed to lack of supplied energy.

Roussos: I gather that Dr Jones's experiments work with isolated curarized muscle. High frequency fatigue can also be due to neuromuscular junction impairment in isolated muscles, or in humans perhaps. Is that due to the same mechanism as you propose for high frequency fatigue at the cell membrane?

Jones: I don't know. Neuromuscular junction failure may be either pre- or postsynaptic—that is, either a reduction in transmitter released or a change of excitability of the postsynaptic membrane. Certainly the changes that occur as the result of high frequency stimulation will make the postsynaptic membrane less susceptible to depolarization by the end-plate potential.

Wiles: Evidence from patients with myasthenia gravis clarifies that slightly. Here the number of functioning receptors is reduced and you do see high frequency fatigue. You also see striking fatigue at very low frequencies of stimulation—in fact, the diagnostic tests for the condition are usually done at about 2 Hz. The analogy between *high* frequency fatigue and neuromuscular transmission block can't be pushed too far, therefore.

Dawson: What is the evidence in these isolated tissues that the high frequency fatigue is *not* related to the metabolic state of the muscle, as opposed to what you have clearly shown, namely that it does correlate with changes in activation?

Jones: The fact that force rapidly returns, on either reducing the stimulation frequency or increasing the stimulation pulse duration, indicates that the original force loss was due to a change in excitability. It is most unlikely that the metabolite concentrations in the muscle would change significantly for the better as a result of a change in stimulation pattern. As I mentioned, the fast muscles do appear to have an additional component to the loss of force which may well be related to metabolite changes.

Edwards: Slowing of the relaxation rate potentiates the frequency–force curve. This can be brought about by cooling and is also observed in patients with hypothyroidism. The relaxation rate is increased in hyperthroidism (Wiles et al 1979). The opposite effect is seen with dantrolene sodium, a muscle relaxant used to treat spasticity. In a patient with spastic paraplegia treated with dantrolene sodium we found a shift of the frequency–force curve to the right (Young & Edwards 1977). This suggested impaired activation, analogous to the low frequency fatigue that we have been discussing.

REFERENCES

Bezanilla F, Caputo C, Gonzalez-Serratos H, Venosa RA 1972 Sodium dependence of the inward spread of activation in isolated twitch muscle fibres of the frog. J Physiol (Lond) 223:507-523

Bigland-Ritchie B, Jones DA, Hosking GP, Edwards RHT 1978 Central and peripheral fatigue in sustained maximum voluntary contractions of human quadriceps muscle. Clin Sci Mol Med 54:609-614

Bigland-Ritchie B, Jones DA, Woods JJ 1979 Excitation frequency and muscle fatigue: electrical responses during human voluntary and stimulated contractions. Exp Neurol 64:414-427

Fink R, Lüttgau HC 1976 An evaluation of the membrane constants and the potassium conductance in metabolically exhausted muscle fibres. J Physiol (Lond) 263:215-238

Grabowski W, Lobsiger EA, Lüttgau HCh 1972 The effect of repetitive stimulation at low frequencies upon electrical and mechanical activity of single muscle fibres. Pflügers Arch Eur J Physiol 334:222-239

Jones DA, Bigland-Ritchie B, Edwards RHT 1979 Excitation frequency and muscle fatigue: mechanical responses during voluntary and stimulated contractions. Exp Neurol 64:401-413

Lüttgau HC 1965 The effect of metabolic inhibitors on the fatigue of the action potential in single muscle fibres. J Physiol (Lond) 178:45-67

Oetliker H 1973 Influence of fatigue on membrane conductance in frog skeletal muscle. Experientia (Basel) 29:747

Wiles CM, Young A, Jones DA, Edwards RHT 1979 Muscle relaxation rate, fibre-type composition and energy turnover in hyper- and hypothyroid patients. Clin Sci (Oxf) 57:375-384

Young A, Edwards RHT 1977 Clinical investigation of muscle contractility. Rheumatol Rehabil 16:231-235

Contractile function and fatigue of the respiratory muscles in man

J. MOXHAM*, C. M. WILES†, D. NEWHAM and R. H. T. EDWARDS

Department of Human Metabolism, University College London School of Medicine, London WC1E 6JJ, UK

Abstract Ventilation depends on the proper functioning of the respiratory muscles. These muscles, like other skeletal muscles, can endure high work loads for only short time periods. The work of breathing in patients with lung disease is increased and it has therefore been argued that respiratory muscle fatigue may develop and contribute to respiratory failure. We have studied the contractile function and fatigue of the sternomastoid muscle and the diaphragm in normal subjects and found that these respiratory muscles have the same contractile properties as limb muscles. If subjected to a high load they develop similar patterns of fatigue. Studies on the sternomastoid muscle in patients with lung disease also confirmed that respiratory stress produces fatigue.

There are few muscles more important than the muscles of respiration; normal ventilation depends on their proper function and if force generation is compromised ventilation can fail. The physiology of these muscles has been extensively reviewed (Campbell et al 1970, Derenne et al 1978, Rochester et al 1979) and it has been proposed that fatigue of the inspiratory muscles may be an important aspect of respiratory failure (Roussos & Macklem 1977, Macklem & Roussos 1977, Edwards 1979). The contractile properties of skeletal muscle can be described in terms of the force response to increasing stimulation frequency (i.e. the frequency: force curve) (Edwards et al 1977a). As a consequence of fatiguing activity this response can be altered, so that less force is generated in response to stimulation at predominantly high or low frequencies (Edwards et al 1977b, Edwards 1978). Using similar techniques to those employed to study limb muscles we have investigated the contractile

---

* *Present address:* The Brompton Hospital, Fulham Road, London SW3 6HP.
† *Present address:* The National Hospital for Nervous Diseases, London WC1N 3BG, UK.

1981 Human muscle fatigue: physiological mechanisms. Pitman Medical, London (Ciba Foundation symposium 82) p 197-212

properties of the muscles of respiration, at rest and in response to stress after fatiguing voluntary activity.

Much of our effort has been directed at an accessory muscle of respiration, the sternomastoid (Moxham et al 1980b). In normal subjects the sternomastoid is active only at high levels of ventilation, serving to raise the sternum and expand the upper thorax. However, in patients with airways obstruction and hyperinflation the sternomastoid is an important inspiratory muscle, often active at rest, and serving to move the whole rib cage upwards during inspiration. Duchenne in 1867 vividly described a young patient with paralysis of virtually all other respiratory muscles who survived several weeks on his sternomastoid muscles only (Duchenne 1959).

Methods

The sternomastoid muscle is amenable to direct study, being flat, thin and superficial, and is easily stimulated with a surface electrode placed over its mid-point, where it receives its innervation from the spinal accessory nerve. When the sternomastoid is stimulated and made to contract the movement produced is lateral flexion of the head and neck to the ipsilateral side and rotation to the contralateral side. Contraction force is in the line of the muscle but direct measurement of this force is not possible. We therefore devised two techniques that recorded alternative force vectors. With the first technique a horizontal component of force in the coronal plane was measured with a strain gauge at the left mastoid process during stimulated contractions of the left sternomastoid. Testing was done with the subject sitting in a dental chair, with adjustable support to the back of the head and neck. Firm pads were applied to the left mastoid region and the right side of the face, holding the head in the anatomical position and reducing movement during stimulation to a minimum. With this arrangement maximum tetanic stimulation caused no head movement, slight elevation of the sternum, and overall muscle shortening amounting to only 3–4% of total muscle length.

With a second technique a force transducer was firmly applied to the sternal tendon of the sternomastoid, displacing it backwards. The subject reclined on a couch at 45° with the head in the anatomical position. When the muscle contracts, the displaced tendon seeks to straighten, and the anterior force vector is recorded. Breathing was suspended at mid-tidal volume during stimulation and the head was fixed by an assistant. To generate the frequency:force curve by either technique the muscle was made to contract using surface electrodes to stimulate the intramuscular nerves, with unidirectional square wave impulses of $50\,\mu s$ duration and 50–80 volts, at increasing frequencies up to 100 Hz. The frequency:force curve was the same whether force was measured at the mastoid or the sternal tendon (Edwards et al 1980). However, recording force at the sternal tendon is a simpler technique, and more acceptable to the subjects being tested, and the equipment is portable.

To complement the sternomastoid studies we have investigated the contractile properties of the diaphragm using a modification of the same basic approach (Moxham et al 1980a). A similar technique, giving comparable results, has also been used by Dr Macklem's group in Montreal. It is not possible to measure directly the force produced by contraction of the diaphragm and the best available index of this force is transdiaphragmatic pressure (Pdi). The right phrenic nerve was stimulated in the root of the neck immediately posterior to the sternomastoid muscle. This causes contraction of the right hemidiaphragm. Gastric pressure (Pg) and oesophageal pressure (Poes) were measured with balloons and Pdi was derived by subtracting Pg from Poes. The electromyogram of the diaphragm was recorded to ensure that phrenic nerve contact was adequate and remained so throughout the stimulation sequence. Two pairs of magnetometers were used to ensure that the configuration of the chest and abdomen (and therefore the diaphragm) were the same before each test. To minimize diaphragmatic shortening, and thereby ensure isometric contraction, the abdomen was strapped and the glottis kept closed during stimulation. To generate a frequency:Pdi curve for the diaphragm the phrenic nerve was stimulated with unidirectional square wave impulses of $100\,\mu s$ duration and 20–50 volts at 1, 10, 20, 50 and 100 Hz.

Results

Study of the sternomastoid muscle and the diaphragm in normal subjects confirms that the contractile properties of these two respiratory muscles are similar to those of other skeletal muscles (Fig. 1). A useful numerical index of the frequency:force relationship is given by the force generated at 20 Hz as a percentage of maximum force. For the sternomastoid, using the mastoid strain gauge technique, in 10 subjects this index was $80.5 \pm 9.2\%$ (mean $\pm$ SD). This compares with $74.2 \pm 7.7\%$ for quadriceps and $73.1 \pm 7.8\%$ for adductor pollicis. For the diaphragm, taking five tests on each of the three normal subjects studied, the Pdi at 20 Hz was $70.0 \pm 7.0\%$ of maximum Pdi.

Failure of force generation at high frequencies—high frequency fatigue—(Edwards 1978) could be an important factor reducing maximum respiratory performance, and is likely to occur in patients with a predisposition to neuromuscular junction failure, as in myasthenia gravis (Fig. 2). This type of fatigue could contribute significantly to the respiratory muscle fatigue and ventilatory failure that can complicate the clinical course of those patients.

During the submaximal contractions of everyday activities the firing frequency of motor neurons is low, between 5 and 30 Hz (Grimby & Hannerz

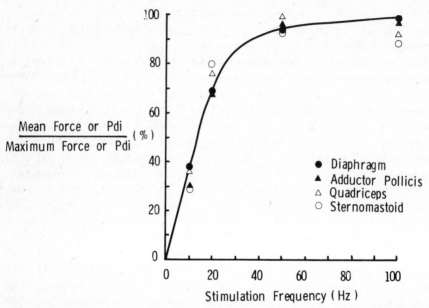

FIG. 1. Frequency: Pdi curve of diaphragm (●) showing mean values for three normal subjects. Also shown are frequency:force data for sternomastoid (○) (*n* = 10), adductor pollicis (▲) (*n* = 10) and quadriceps (△) (*n* = 10). In some subjects the maximum force was produced by stimulation at 50 Hz rather than 100 Hz; the mean force values do not therefore reach 100%. (From Moxham et al 1981 by permission of the Editor of *Thorax*.)

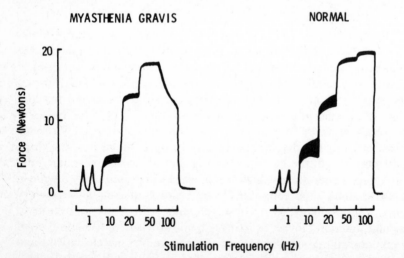

FIG. 2. Frequency:force stimulation myogram of the sternomastoid recorded from the sternal tendon. The response of the fresh muscle in a normal subject is shown on the right. Left, the result in a patient with myasthenia gravis, illustrating high frequency fatigue.

1977), and even with sustained maximum contractions the firing frequency is only greater than 30 Hz for a few seconds (Jones et al 1979). Thus, for the most part motor neuron firing frequency is on the steep part of the frequency:force curve. It would therefore be particularly important if force generation at these low frequencies was reduced—that is, if low frequency fatigue were to develop (Edwards et al 1977b).

To investigate the effect of increased work on the contractile properties of the diaphragm we asked normal subjects to undertake a period of breathing through an inspiratory resistance. An inspiratory resistance was selected for each subject so that 70% of maximum transdiaphragmatic pressure was generated with each breath for as long as possible, and loaded inspiration was stopped when this target could not be reached for three out of four breaths. Each subject was studied on more than one occasion and the time for which the target could be sustained varied from one test to the next and between subjects, ranging from 10 to 20 minutes. After the subject had rested for 10 minutes the contractile properties of the diaphragm were re-examined. The contractile response of the diaphragm after fatigue was reduced at low frequencies but relatively well maintained at high frequencies (Fig. 3). The mean Pdi at 20 Hz as a percentage of maximum Pdi declined from $69.8 \pm 3.8\%$ (mean $\pm$ SD) in fresh muscle to $49.7 \pm 7.0\%$ ($n = 15$, $P < 0.01$) after fatigue. Muscle excitation, however, remained normal, as judged by the amplitude of the evoked muscle action potential.

We did similar experiments on the sternomastoid. Firstly, normal subjects breathed through inspiratory resistances in such a way that 70% of maximum inspiratory mouth pressure was developed with each breath and this target was sustained for 200 breaths. In a second series of experiments, normal subjects undertook sustained maximum voluntary ventilation (MVV) for 10 minutes, with end-tidal CO_2 kept constant and without hypoxia.

In seven respiratory loading tests the force at 20 Hz fell from 80% of maximum to 38%, and for six sustained MVV tests from 82.5% to 55.6%. It is therefore clear that the contractile properties of the respiratory muscles—the sternomastoid and diaphragm—are the same as those of other skeletal muscles, and that with respiratory stress both muscles can develop low frequency fatigue.

Because the work of breathing is greatly increased in patients with lung disease and because high loads have to be sustained by the muscles of inspiration for prolonged periods, without rest, it seems likely that low frequency fatigue could well develop. The recording of frequency:force curves from the sternal tendon of sternomastoid is a particularly useful technique because it is acceptable to patients.

We have studied a small group of patients with chronic airways obstruction. The patients were not ill at the time of the study, being well enough to walk to

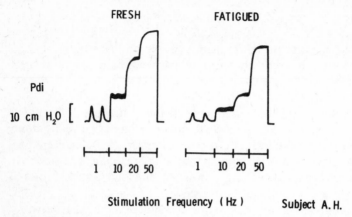

FIG. 3. Diaphragmatic Pdi in response to right phrenic nerve stimulation at 1, 10, 20 and 50 Hz. On the left is the result in fresh muscle. Right, the result 10 min after inspiratory loading, illustrating low frequency fatigue. (Subject A.H.)

the respiratory laboratory. The sternomastoid frequency:force curves on seven such patients were entirely normal with a force at 20 Hz of $78 \pm 6.4\%$ (mean $\pm$ SD) of maximum force. To investigate the contractile properties after respiratory stress we asked several of the patients to undergo a 10-minute period of maximum voluntary ventilation, keeping end-tidal CO_2 steady and adding oxygen to avoid hypoxia. Fig. 4 shows the contractile properties of the sternomastoid in a lady, aged 60, with severe airways obstruction, with a resting minute ventilation of seven litres and a forced expiratory volume in one second (FEV_1) of only one-third of a litre. As would be expected the level of sustained voluntary ventilation was extremely low, the mean minute value over the ten minutes of the test being 13 litres. The closed circles in Fig. 4 indicate a normal frequency:force curve before the respiratory stress. The open circles show that 10 minutes after the end of the sustained MVV substantial low frequency fatigue was present.

In four patients with airways obstruction—with a range of impaired ventilation from 15% to 80% of normal—sustained MVV for 10 min produced significant low frequency fatigue, the force at 20 Hz falling from a mean value of 81% of maximum to 55% ($P < 0.02$).

If a particular set of respiratory muscles bear an increased proportion of the work of respiration because of, for example, wasting and atrophy of other muscles, they are likely to develop fatigue, as the following case report indicates.

A man of 44 had had poliomyelitis at the age of 23, affecting many of his respiratory muscles with relative preservation of the sternomastoids. There was severe wasting of the intercostal muscles and paradoxical abdominal

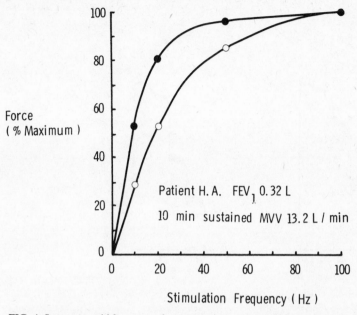

FIG. 4. Sternomastoid frequency:force curve in a patient with chronic obstructive airways disease (Patient H.A. FEV_1, 0.32 l, 10 min sustained maximum voluntary ventilation, 13.2 l/min). ● : fresh muscle. ○ : after 10 min of maximum voluntary ventilation.

movement, indicating diaphragmatic paralysis or weakness. When he stood up the sternomastoid muscles were barely active but when he lay flat they became very active. The vital capacity was 1.4 litres standing and 0.9 litres supine. When he was awake and upright the arterial oxygen tension was 8.2 kPa and Pco_2 was 8.7. After he had lain flat for several minutes the Po_2 fell to 6.3 kPa and the Pco_2 rose to 9.3 kPa. The patient was not able to remain asleep for more than half an hour. The sternomastoid frequency:force curve was recorded after the patient had remained upright all day: the patient then lay flat for one hour and after an additional 10 min of sitting upright to allow breathlessness to subside we re-tested the sternomastoid (Fig. 5). The stress of lying flat for one hour produced substantial low frequency fatigue, the force at 20 Hz falling from 73% to 44% of maximum.

Conclusion

Respiratory muscles develop low frequency fatigue when subjected to a sufficient loading. Because the frequency:force curve is shifted to the right it becomes necessary either to increase stimulation frequency or recruit addi-

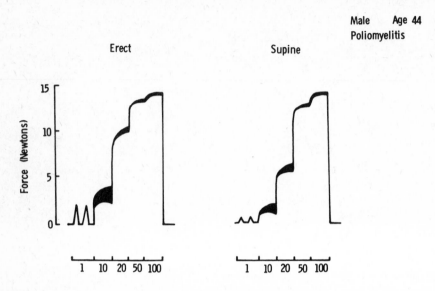

FIG. 5. Sternomastoid stimulation myogram in a patient with respiratory muscle weakness with sparing of sternomastoid muscles. Left, contractile response when patient was standing and sternomastoid not active. Right, response after one hour of lying supine, associated with intense sternomastoid contractions. The response on the left is normal and that on the right shows low frequency fatigue.

tional units if force is to be maintained, and the central respiratory drive must be increased if ventilation is not to fall. As with other muscles (Jones et al 1979) there is presumably an upper limit of motor neuron firing frequency that can be sustained for more than a few seconds, and with the development of low frequency fatigue it is implicit that the maximum sustained level of ventilation must fall. For many patients it is this maximum sustained level of ventilation that limits activity (Clark et al 1969). A further consequence of low frequency fatigue could be that maximal respiratory drive (i.e. maximal firing frequency) becomes necessary to maintain ventilation and thus any depression of drive (by sedatives, CO_2 narcosis, or excess O_2 therapy) would cause a fall in ventilation. The long duration of this type of fatigue (Edwards et al 1977b, Moxham et al 1980b) suggests that in some patients one of the benefits of assisted ventilation may be that the respiratory muscles are rested and their contractile properties can return to normal. Therapy that increases respiratory drive will improve force generation and if drugs that specifically reverse the changes of low frequency fatigue can be developed they could become an important part of the overall treatment of respiratory failure.

Acknowledgements

The authors gratefully appreciate the help and advice of Drs S. G. Spiro, S. Freedman, M. Green and A. J. R. Morris, and the technical assistance of Mr A. Cobley. Support from The Wellcome Trust and the Muscular Dystrophy Group of Great Britain is gratefully acknowledged.

REFERENCES

Campbell EJM, Agostoni E, Newsom Davis J 1970 The respiratory muscles: mechanics and neural control, 2nd edn. Lloyd-Luke, London

Clark TJH, Freedman S, Campbell EJM, Winn RR 1969 The ventilatory capacity of patients with chronic airways obstruction. Clin Sci Mol Med 36:307-316

Derenne JP, Macklem PT, Roussos C 1978 The respiratory muscles: mechanics, control and pathophysiology. Am Rev Respir Dis 118:119-33, 373-90, 581-601

Duchenne GB 1959 Physiologie des mouvements demontrée a l'aide de l'expérimentation éléctrique et de l'observation clinique et applicable à l'étude des paralysis et defermation, 1867. Kaplan EB (transl) English translation: Physiology of motion. Saunders, Philadelphia

Edwards RHT 1978 Physiological analysis of skeletal muscle weakness and fatigue. Clin Sci Mol Med 65:1-8

Edwards RHT 1979 The diaphragm as a muscle. Mechanisms underlying fatigue. Am Rev Respir Dis 119:81-84

Edwards RHT, Young A, Hosking GP, Jones DA 1977a Human skeletal muscle function: description of tests and normal values. Clin Sci Mol Med 52:283-290

Edwards RHT, Hill DK, Jones DA, Merton PA 1977b Fatigue of long duration in human skeletal muscle after exercise. J Physiol (Lond) 272:769-778

Edwards RHT, Moxham J, Newham D, Wiles CM 1980 Alternative techniques for recording the frequency:force curve of the sternomastoid muscle in man. J Physiol (Lond) 305:4P-5P

Grimby G, Hannerz J 1977 Firing rate and recruitment order of toe extensor motor units in different modes of voluntary contraction. J Physiol (Lond) 241:45-57

Jones DA, Bigland-Ritchie B, Edwards RHT 1979 Excitation frequency and muscle fatigue: mechanical responses during voluntary and stimulated contractions. Exp Neurol 64:401-413

Macklem PT, Roussos CS 1977 Respiratory muscle fatigue: a cause of respiratory failure? Clin Sci Mol Med 53:419-422

Moxham J, Morris AJR, Spiro S, Edwards RHT, Green M 1980a The contractile properties and fatigue of the diaphragm in man. Clin Sci Mol Med 58:6P

Moxham J, Wiles CM, Newham D, Edwards RHT 1980b Sternomastoid muscle function and fatigue in man. Clin Sci Mol Med 59:463-468

Moxham J, Morris AJR, Spiro SG, Edwards RHT, Green M 1981 Contractile properties and fatigue of the diaphragm in man. Thorax, in press

Rochester DF, Arora NS, Braun NMT, Goldberg SK 1979 The respiratory muscles in chronic obstructive pulmonary disease (COPD). Bull Eur Physiopathol Respir 15:951-975

Roussos CS, Macklem PT 1977 Diaphragmatic fatigue in man. J Appl Physiol 43:189-197

DISCUSSION

Macklem: Is there a certain level of rest necessary for recovery from fatigue? If you fatigue a muscle and put it to rest, it recovers. If a contraction is sustained, however, does it take longer to recover, or does it even not recover at all? The question is potentially an important one ethically in experiments with patients. We have done the same sort of things as you have, Dr Moxham. What worries me is that if a patient has to breathe continually against a high load and you deliberately fatigue his inspiratory muscles, he may never recover.

Moxham: The cases we studied did recover, presumably because the work load on the respiratory muscles at rest was much less than during sustained maximum voluntary ventilation or when subjects are breathing through an inspiratory resistance.

Macklem: Our patients recovered too, but I have seen patients who have developed a form of chronic inspiratory muscle fatigue, in whom artificial ventilation has improved their respiratory function to a much higher level than it had been at for weeks previously. So a state of chronic fatigue may exist. Although we have got away with our experiments so far, sooner or later, if we continue to fatigue the inspiratory muscles of patients deliberately, we shall have trouble.

Edwards: In terms of the time course of recovery of force, there are precedents. The recovery of maximum force was studied by Lind's group after hand-grip contractions (Lind & Petrofsky 1979). Recovery was rapid, in minutes. Eric Hultman's studies (this volume, p 19-35) indicate that recovery has a half-time of about 30 seconds, a bit slower than phosphocreatine resynthesis. That is true for high frequency fatigue. We have seen that this may be, to some extent, attributed to recovery of the action potential (Wiles et al, this volume, p 264-277). But our low frequency fatigue studies showed slow recovery, over 24 hours. You have raised the important point that the person is continuing to exercise (i.e., to breathe) during the period when such recovery might be looked for.

Roussos: In the training of the respiratory muscles, how much training should be given to avoid low frequency fatigue, when the respiratory muscles are already working against a resistance? Are there any guidelines? Presumably the development of low frequency fatigue should be avoided, particularly in patients with chronic obstructive pulmonary disease, so we need to know how much training we can give these patients.

Edwards: It may be too late to do this in a patient with respiratory problems, because you may be unable to improve his function without simultaneously running into problems with fatigue.

Roussos: John Moxham showed that patients with chronic obstructive pulmonary disease didn't have chronic fatigue of the sternomastoid muscle.

Moxham: That is true initially, but in the process of trying to train them, they may develop fatigue sooner than you can cope with it. So you may get harmful rather than beneficial effects of training.

Macklem: The accumulating experience of a number of those who are attempting to train the inspiratory muscles in patients with chronic airways obstruction is that some patients clearly can be trained. They improve, by a number of different indices. Another group of patients are untrainable, possibly because they are already completely trained or because they are in a state of chronic fatigue and you can't do anything with them.

Edwards: The question would be whether a test of the nature we have heard about might help to distinguish those groups.

Stephens: Have you taken the opportunity in your patients, Dr Moxham, simply to record the surface EMG from the sternomastoid muscle? For example, in the young man who fatigued in one hour, just by lying down, does the EMG get progressively bigger in his sternomastoid muscle?

Moxham: We have not done that experiment. We have studied the smoothed rectified EMG (SREMG) in several muscles when a state of low frequency fatigue is present. With fatigue there is a shift in the force: SREMG curve; for a given force, more SREMG activity is required. You can also show this with ventilation experiments. If you set the subject a particular target, say to generate 50% of his maximum inspiratory mouth pressure, there is more EMG activity from the sternomastoid when low frequency fatigue is present than in the fresh muscle. This supports the idea that with fatigue more respiratory drive is required. We have not recorded the EMG actually during the respiratory stress that produces fatigue.

Macklem: The problem of looking just at the amplitude of the EMG is that the force developed may vary from breath to breath. The amplitude record alone is not enough. You need an index of fatigue that is independent of force, which is why we have turned to the power spectral shift.

Donald: Have you any quantitative evidence of hypertrophy in the sternomastoid muscles, in severe obstructive respiratory disease?

Moxham: Respiratory muscles have been measured (weight and fibre size) at *post mortem*, but this is complicated because the muscle mass is greatly affected by nutrition. The diaphragm shares in the general atrophy of all muscles that is such a striking feature of severely ill or terminally ill patients (Rochester et al 1979). I don't know of any measurements specifically on the sternomastoid muscles of patients with respiratory disease.

Donald: It is curious that whereas heart muscle hypertrophies when it is overloaded and does so quite rapidly, in weeks or months, respiratory muscle does not obviously hypertrophy. It may do so, but it has been difficult to demonstrate.

Macklem: Hypertrophy results primarily from isometric contractions. Isotonic contraction doesn't lead to hypertrophy. Marathon runners don't get hypertrophied leg muscles, but weight lifters develop hypertrophied arm muscles! In particular conditions the diaphragm contracts quasi-isometrically, but it usually contracts isotonically.

Donald: There is a lot of isometric holding in certain parts of the thorax.

Macklem: Perhaps the parasternal muscles contract isometrically.

Roussos: There is one study of the accessory muscles of respiration in patients with chronic obstructive pulmonary disease in which hypertrophy of all the accessory muscles was indicated (Thompson et al 1964).

Campbell: I think the attention paid to the sternomastoid muscle by clinicians is rather exaggerated. It is nothing like as commonly in use in chest disease as is usually thought. The physical prominence of the sternomastoid is largely due to the sculpturing of the soft tissues. The muscle often appears to stand out when it is not active. But this comment doesn't deny the importance of what you have found, because it is a muscle that one may be called on to use in circumstances when fatigue would be undesirable.

Moxham: I would entirely agree with that view. It is very odd that some patients seem to use their accessory muscles a lot and others do not. In the studies we did, the patients used them vigorously.

Campbell: The most common clinical situation where fatigue of the sternomastoid is important may be in the asthmatic patient whose chest is inflated. I think that fatigue of the sternomastoids may contribute to the distress of these patients.

Edwards: Let me make a distinction here between different forms of EMG changes, because we shall be going on to consider power spectral changes. Olof Lippold was one of the first to talk about EMG changes with fatigue (Edwards & Lippold 1956). This was a change *while* the muscle was contracting and the fatiguing process was taking place. There was a greater amplitude of the smoothed rectified EMG while the muscle was contracting submaximally. We have to distinguish this from what John Moxham described, that 10 minutes after the fatiguing activity, and for a long time afterwards in recovery, there was increased electrical activity (SREMG). The contractions studied by Dr Lippold and ourselves were submaximal, and this needs to be emphasized. With a sustained maximal contraction the smoothed rectified EMG is reduced, as shown here by Brenda Ritchie (this volume, p 130-148).

Donald: The assisted ventilation of severe obstructive respiratory disease is being used less now than formerly, so apparently this resting of the muscles has not been such a success as people expected. It is sometimes difficult to get such patients off the ventilator; there doesn't appear to be the degree of recovery that one hoped for. But there are other complications. Such patients are severely obstructed, with secretions, and areas of collapse can occur.

Moxham: There are numerous factors in respiratory failure and in patients who reach the stage of needing assisted ventilation. I am simply suggesting that one beneficial effect could be that it allows the respiratory muscles to recover from this sort of fatigue.

Donald: You mentioned how helpful it would be if the respiratory muscles could be given fast-frequency stimulation in this fatigued state. Is there any way of doing that with drugs?

Moxham: I said that, faced with the fact that at low stimulation frequencies less force is generated, you need to increase the motor neuron firing frequency, and thus to increase respiratory drive. I don't know whether respiratory stimulants do this. Perhaps the main point is at least to avoid depressing respiratory drive. There are patients in respiratory failure where even allowing them to sleep is often enough to tip them over the edge, and keeping them awake and alert can be vital. This is one of the benefits of frequent physiotherapy.

Edwards: Maintaining wakefulness is what Moran Campbell used to call 'controlled brutality'! May I bring us back to physiology? Another feature of respiratory muscles that is interesting and possibly unique is the fact that as the mechanics of the chest change and the lungs become larger, the respiratory muscles become much shorter. We have started to investigate, with John Moxham, the phenomenon originally shown by Rack & Westbury (1969) in the cat soleus muscle and known as the 'length dependence of activation'. As muscle is allowed to shorten, the force generated is not only less, as shown by A. V. Hill (1970) in conditions of maximal activation; but disproportionately less force is generated at low frequencies than at high frequencies. If, in addition to low frequency fatigue, the muscle is also shortened, the phenomenon of length dependence of activation is likely to have implications for respiratory failure (Edwards 1979).

Campbell: Dr Moxham studied the effects of breathing through an inspiratory resistance on the development of fatigue in the diaphragms of normal subjects. Do you know whether you can produce low frequency fatigue in patients with bad chests by these manoeuvres?

Moxham: No; we haven't studied the diaphragm in patients, but the experiments showing fatigue of the diaphragm in normal subjects and fatigue of the sternomastoid in patients strongly suggest that similar fatigue is likely to develop in the diaphragm of patients with severe lung disease.

Campbell: My feeling is that, were patients to do the MVC type of manoeuvre, they would very likely develop much more force in the diaphragm than is needed for the MVC, so the fact that they are developing that force at

MVC is not an argument that such force is needed for MVC. The chest is so mechanically bad that much of the force will not contribute to ventilation. I wonder if there may be an important difference between a muscle that is occasionally used, as against a muscle that is used all the time. I can see why low frequency fatigue may be important for the sternomastoid muscle when it *is* used and that this type of fatigue may have unfortunate effects, but is there low frequency fatigue in the diaphragm in patients?

Moxham: As I say, such fatigue of the diaphragm is very likely; it would, of course, be very nice to demonstrate it unequivocally.

Roussos: We asked patients with CO_2 retention to hyperventilate in order to decrease the CO_2 level by 6 mmHg; this meant increasing minute ventilation by 5–7 litres, depending on the patient. We studied the force–frequency curve in three of these patients and they developed low frequency fatigue of the diaphragm.

Campbell: That is not quite the answer to my question. Did you also measure transdiaphragmatic pressures?

Roussos: Yes. They were developing pressures which were 50% of the maximum.

Donald: We had an idea some years ago that if failure of the myocardium was helped by digitalis, failure of a somatic muscle such as the respiratory muscle might also be helped by this drug. Unfortunately, digitalis doesn't help exhausted skeletal muscle (dynamic or isometric exercise) in any way (Bruce et al 1968).

Edwards: Another possible benefit of a period of assisted ventilation is that it might allow the mechanics of the chest to return towards normal, by alterations in fibre length consequent on a reduction in lung volume.

Donald: This is why assisted ventilation often goes so wrong, because it can be even more inefficient than the diseased breathing the patients already have. Almost any type of artificial respiration is to some extent inefficient, because it is being performed by positive and not negative pressure, and therefore often the lung is not working so well as previously, particularly as regards oxygenation.

Macklem: What do you mean by 'inefficient'?

Donald: You get collapse in certain areas of the lung and much more expansion in some other areas.

Macklem: You can certainly improve CO_2 exchange by artificial ventilation. Rochester & Braun (1979) showed that if patients with chronic lung disease are put into a tank respirator the EMG of their respiratory muscles falls to zero; they relax their muscles entirely. So if that's the object, it's successful.

Newsholme: May I make a metabolic statement and a suggestion? Dr Pugh asked me earlier about ketone bodies as a fuel for muscles, and I generally discounted the idea from the evidence from running and similar sorts of exercise (p 100-101). However, we did a comparative study of the enzymes involved in fuel utilization (Beis et al 1980). An intriguing aspect was that if we categorized certain muscles as having a vital function, such as the respiratory muscles, heart muscle, some of the intestinal muscles, these muscles have very high levels of enzymes capable of using ketone bodies. To take a simple example, the pectoral muscle of the pigeon, which can fly several hundred miles without stopping, contains large amounts of fatty acid-oxidizing enzymes and low levels of ketone body enzymes, whereas in the gizzard, not usually considered to be an 'important' muscle, the activities of ketone body enzymes are very high, 10-fold higher than in the quadriceps muscle of man, for example. We also find high activities of ketone body enzymes in postural muscles and in the diaphragm. We rationalize this physiologically by suggesting that in situations in which there may be severe hypoglycaemia, such as in prolonged starvation, severe exercise or imbalance of hormones, the blood concentrations of ketone bodies increase to high levels, and these vital muscles could then use ketone bodies instead of glucose. I therefore suggest that in acute respiratory distress, infusion of ketone bodies, or administration of diets containing short-chain fatty acids, which would favour ketone body formation, might help.

Edwards: On another point, an advantage of constructing frequency–force curves is that they provide another way of looking for factors influencing fatigue, and in particular a way of studying the pharmacology of skeletal muscle. John Moxham mentioned that respiratory alkalaemia can reverse low frequency fatigue. More recently Ms Sandra Howell from McGill University has been working with David Jones, looking at the effects of caffeine and other compounds on low frequency fatigue.

Jones: In my talk I tried to indicate that the underlying cause of low frequency fatigue is the change in amplitude of the twitch; it becomes smaller and therefore the force generated at low frequencies is less. Anything that reverses this change in the twitch will most likely cause more force to be generated at low frequencies of stimulation. A number of compounds are known to potentiate the skeletal muscle twitch, including caffeine, other methylxanthines, zinc, and thiocyanate (Sandow 1965).

With Sandra Howell from Professor Roussos' laboratory we have tested a number of the methylxanthines and found that caffeine and related compounds do indeed reverse the effects of low frequency fatigue in a variety of isolated mammalial muscle preparations. If an effective but safe dose can be found, this may become a useful way of managing this type of fatigue in patients.

REFERENCES

Beis A, Zammet VA, Newsholme EA 1980 Activities of 3-hydroxybutyrate dehydrogenase, 3-oxoacid CoA transferase and acetoacetyl-CoA thiolase in relation to ketone body utilisation in muscle from vertebrates and invertebrates. Eur J Biochem 104:208-215

Bruce RA, Lind AR, Franklin D, Muir AL, Macdonald HR, NcNicol GW, Donald KW 1968 The effects of digoxin on fatiguing static and dynamic exercise. Clin Sci (Oxf) 34:29-42

Edwards RHT 1979 The diaphragm as a muscle. Mechanisms underlying fatigue. Am Rev Respir Dis 119:81-84

Edwards RG, Lippold OCJ 1956 The relation between force and integrated electrical activity in fatigued muscle. J Physiol (Lond) 132:677-681

Hill AV 1970 First and last experiments in muscle mechanics. Cambridge University Press, London, p 126

Lind AR, Petrofsky JS 1979 Amplitude of the surface electromyogram during fatiguing isometric contractions. Muscle Nerve 2:257-264

Rack PMH, Westbury DR 1969 The effects of length and stimulus rate on tension in the isometric cat soleus muscle. J Physiol (Lond) 204:443-460

Rochester DF, Braun NMT 1979 The diaphragm and dyspnea: evidence from inhibiting diaphragmatic activity with respirators. Am Rev Respir Dis 119: part 2, 77-80

Rochester DF, Arora NS, Braun NMT, Goldberg SK 1979 The respiratory muscles in chronic obstructive pulmonary disease (COPD). Bull Eur Physiopathol Respir 15:951-975

Sandow A 1965 Excitation-contraction coupling in skeletal muscle. Pharmacol Rev 17:265-320

Thompson WT, Patterson JL, Shapiro W 1964 Observations on the scalene respiratory muscles. Arch Intern Med 113:856-865

Wiles CM, Jones DA, Edwards RHT 1981 Fatigue in human metabolic myopathy. This volume p 264-277

Neural drive and electromechanical alterations in the fatiguing diaphragm

C. ROUSSOS and M. AUBIER

Meakins Christie Laboratories, McGill University Clinic, Royal Victoria Hospital, Montreal, Canada H3A 2B4

Abstract It is suggested that respiratory failure in the compromised circulation might occur as a result of respiratory muscle fatigue in the presence of adequate neural drive and muscle excitation. As the cardiac output decreases acidosis develops and ventilation increases, resulting in an increase in the work of breathing, which requires the delivery of large supplies of energy. As these demands cannot be met by the energy supply, because of low cardiac output, the diaphragm fails as a force generator and respiratory failure ensues. Diaphragmatic fatigue may occur in normal subjects if the pressure developed with each breath is greater than 40% of the maximum transdiaphragmatic pressure and hypoxia predisposes the diaphragm to fatigue. Diaphragmatic fatigue, as in other skeletal muscles, might be located either at the neuromuscular junction or distal to it and can be detected either by phrenic stimulation or by frequency analysis of the myoelectric signal. Phrenic stimulation shows that after fatigue the diaphragm develops less force at any frequency of stimulation, but the loss of force at low frequencies persists for a longer period than at high frequencies. Frequency analysis of the electromyogram reveals that the power spectrum shifts to lower frequencies. This shift occurs long before the diaphragm fails as a force generator.

Does the diaphragm become fatigued? Does it fail the same way as the heart and other skeletal muscles fail under stress? Although these questions were asked by Hippocrates and have been asked by many others since, respiratory muscle fatigue has only recently been demonstrated to be a potential cause of respiratory failure (Bradley & Leith 1978, Roussos & Macklem 1977, Roussos et al 1979). Respiratory muscle failure can occur as a result of any derangement involving central mechanisms, peripheral neuromuscular junctions and contractile components. An example of failure of the central nervous system to provide adequate drive to respiratory muscle is provided by patients in respiratory failure who 'won't breathe'; failure of the neuromuscular junction or any process beyond it is seen in patients who 'cannot breathe'.

1981 Human muscle fatigue: physiological mechanisms. Pitman Medical, London (Ciba Foundation symposium 82) p 213-233

213

Respiratory muscle failure resulting from fatigue of the respiratory muscles is analogous to failure in other skeletal muscles, both central and peripheral. Although central fatigue may be an important factor in maintaining normal ventilatory function, our investigations suggest peripheral muscle fatigue as the predominant derangement in failure to generate an adequate pleural pressure. Respiratory muscles, like other skeletal muscles, consume chemical energy and produce work; thus they can be likened to an engine. It may be hypothesized therefore that respiratory muscle fatigue (as opposed to central or neuromuscular transmission fatigue) occurs when the rate of energy consumption exceeds the rate at which energy is supplied. Factors that determine the energy demands of the inspiratory muscles include the work of breathing (minute ventilation, frequency and tidal volume, compliance and resistance), strength (lung volume, atrophy, prematurity, neuromuscular disease, nutritional status) and efficiency (resistance, lung volume). Factors that determine the energy available to the inspiratory muscles include the oxygen content of arterial blood (O_2 saturation, haemoglobin concentration), the inspiratory muscle blood flow (cardiac output, distribution of perfusion, force of inspiratory muscle contraction), substrate concentrations in blood, energy stores in the muscles, and the ability to extract the energy sources. We shall discuss here some of the factors that we have studied, as well as some characteristics of the failing diaphragm in the presence of adequate neural drive.

Diaphragmatic fatigue in an animal model

We have developed an animal model in which dogs are submitted to cardiogenic shock through pericardial tamponade (Aubier et al 1980). Twelve dogs were used. A polyvinyl catheter was introduced into the pericardium through the fourth left intercostal space and shock was induced by injecting warm saline solution (37 °C) through the catheter. Cardiac output was decreased in each dog by $70 \pm 5\%$ of control values and was then maintained at that level throughout the experiment; this was achieved by monitoring blood pressure, which averaged 58 ± 5 mmHg for the group, and by measuring the mixed venous Po_2, and the cardiac output by the Fick equation every 10 min. Corrections were made for the amount of pericardial fluid present.

All animals gradually increased their ventilation (Fig. 1), reaching a peak in about an hour. Ventilation then decreased and the dogs died from respiratory failure $2-2\frac{1}{2}$ hours from the beginning of the experiment. Respiratory failure could have resulted from changes in the mechanical properties of the respiratory system. However, Table 1 shows that neither lung resistance nor lung compliance changed during the experiment. Furthermore, in four dogs

TABLE 1 Mean values ± SEM of functional residual capacity (FRC), total pulmonary resistance and dynamic compliance before and during cardiogenic shock in 12 dogs

	Control value before shock	60 min shock	120 min shock	140 min shock
FRC (ml)	900 ± 54	909 ± 46	896 ± 45	898 ± 51
Total pulmonary resistance (cmH$_2$0 l^{-1} s^{-1})	3.54 ± 0.46	3.72 ± 0.56	3.64 ± 0.50	3.56 ± 0.44
Dynamic compliance (1/cmH$_2$O)	0.054 ± 0.015	0.064 ± 0.019	0.068 ± 0.019	0.066 ± 0.017

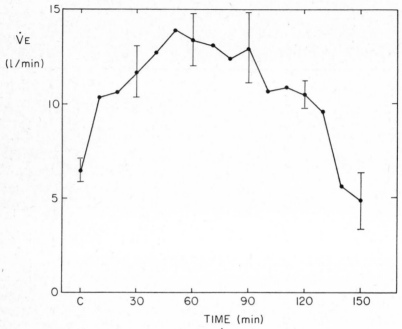

FIG. 1. Mean Changes in minute ventilation ($\dot{V}_E$) during cardiogenic shock. C, control value before shock. Bars, standard error (SEM).

that were studied in a body plethysmograph, the functional residual capacity (FRC) remained unchanged. Therefore gross changes in the length and geometry of the respiratory muscles cannot account for the decrease in ventilation and respiratory failure. We also monitored gastric pressure (Pg) and found that the end-expiratory Pg remained unchanged during the run, therefore excluding large geometrical alterations of the diaphragm at constant volume.

A decrease in ventilation might be brought about by a decrease in central neural drive. However, Fig. 2 does not support this. Tidal volume (V_T), transdiaphragmatic pressure (Pdi) and tracheal pressure during inspiratory effort against closed airways (P_T) (Whitelaw et al 1975) simultaneously increased and then decreased. In contrast, the integrated phrenic nerve activity (Ephr) continued to increase until the last breath. The mean control value and values at 60 min and 140 min were: minute ventilation, 6.5 ± 0.6, 13.5 ± 0.3 and 4.9 ± 1.5 l/min; Pdi, 7.9 ± 1, 12 ± 2, 4.2 ± 1 cmH$_2$O; pressure generated 0.3 s after the beginning of inspiration against closed airways ($P_{0.3}^0$), 3.9 ± 0.5, 7.5 ± 1, 2.5 ± 0.6 cmH$_2$O; and Ephr, 10, 15, 29 arbitrary units. From these results we can conclude that central respiratory drive is more than

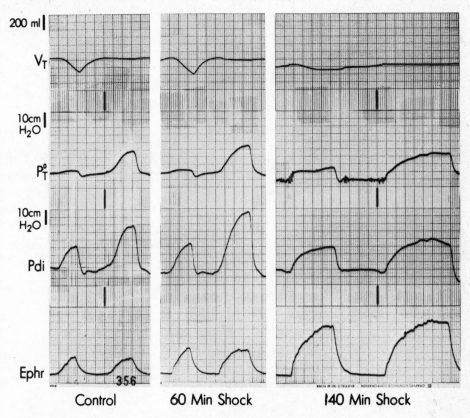

FIG. 2. Typical record of one dog showing the changes in tidal volume (V_T, first trace), tracheal pressure (P_T^0, second trace), transdiaphragmatic pressure (Pdi, third trace) and electrical integrated activity of the phrenic nerve (Ephr, fourth trace) during inspiratory effort against closed airways and during the preceding breath. Left panel: during control period. Middle panel: 60 min after the onset of cardiogenic shock. Right panel: 140 min after the onset of cardiogenic shock, before death.

adequate. Thus factors in the periphery must be responsible for the respiratory failure. As discussed above, these factors do not involve changes in lung and chest wall mechanics. Derangements in the functioning of the neuromuscular junctions or in the contractile machinery of the diaphragm are other possible causes of respiratory failure which could explain our results. The neuromuscular junction is not obviously responsible for the diaphragmatic failure. Fig. 3 shows that the integrated diaphragmatic electromyogram (Edi) continued to increase in parallel with the electrical activity of the phrenic nerve. In fact the relationship between the Edi and Ephr remained constant throughout the experiment in all animals ($r = 0.83$). Figs. 2 and 3

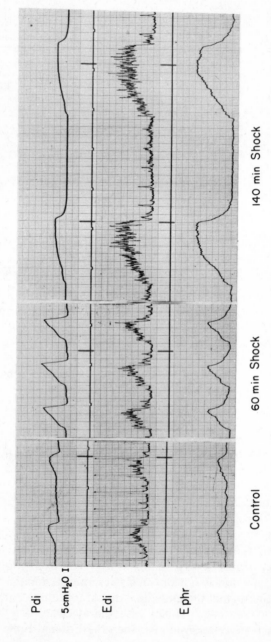

FIG. 3. Typical record of one dog showing the evolution during cardiogenic shock of transdiaphragmatic pressure (Pdi, upper trace), electrical integrated activity of the diaphragm (Edi, middle trace), and electrical integrated activity of the phrenic nerve (Ephr, lower trace). Left panel: during control period. Middle panel: 60 min after the onset of cardiogenic shock. Right panel: 140 min after the onset of cardiogenic shock, before death. Note that Edi and Ephr continued to increase until the end, while Pdi fell after an increase. The decrease in the size of the EKG artifact on the Edi trace is a consequence of the injection of saline into the pericardium.

therefore demonstrate a lack of association between nerve drive, muscle excitation and muscle contraction.

Thus in these experiments we have shown that when the cardiac output is decreased, the main respiratory muscle fails. This failure is due to factors that are primarily beyond the neuromuscular junction.

What is the underlying mechanism of diaphragmatic failure? We suggested earlier that a decrease in cardiac output will result in a decrease in diaphragmatic blood flow. Our preliminary results using radionuclide microspheres (Rudolph & Heymann 1967) do not yet provide a satisfactory answer on this. Despite shock we find that the working diaphragm at the peak of ventilation increases its blood flow. However, we have not determined whether this increase is sufficient to meet the energy demands. From our biochemical analysis, the glycogen content of the diaphragm is considerably depleted while the lactate content is greatly increased. Fig. 4 shows that for the whole group of dogs glycogen decreased from 1.83 ± 0.3 mg/g tissue to 0.54 ± 0.33 mg/g (mean $\pm$ SEM) and lactate in diaphragmatic muscle increased from 2.83 ± 19 mmol/kg to 5.10 ± 0.53 mmol/kg, while blood lactate concentrations rose from 1.08 ± 15 mmol/l to 9.48 ± 2.7 mmol/l. The levels of muscle glycogen cannot account for the muscle failure since glycogen was not completely depleted in any of the dogs. The increase in muscle lactate may partly explain our results. Fuchs et al (1970) found that an increase in H^+ concentration interferes with the binding of Ca^{2+} to troponin by lowering the apparent binding constant. In another study, Nakamura & Schwartz (1972) found that a decrease in pH increased the Ca^{2+}-binding capacity of the sarcoplasmic reticulum. Both mechanisms would lead to a decrease in the number of calcium ions bound to troponin during excitation–contraction coupling. This would reduce the active interaction between actin and myosin and thus decrease the contractile force (Fitts & Holloszy 1976).

Our recent results on the effect of aminophylline on the diaphragm may be explicable in this context. We have found that aminophylline improves diaphragmatic performance. If the Ca^{2+} concentration is reduced, thereby impairing the contraction, aminophylline may help by increasing the release of Ca^{2+} from the sarcoplasmic reticulum. In addition, phosphofructokinase activity is inhibited by a decrease in pH (Trivedi & Danforth 1966). By this means, the accumulation of lactic acid could inhibit glycolysis. This pathway is the only one known to produce the ATP necessary for muscle contraction when the muscles work anaerobically. In our model, blood flow to the diaphragm might be inadequate, thereby limiting aerobic metabolism. The accumulation of lactate increases hydrogen ion concentration which in turn impairs ATP production via the anaerobic pathway. Thus the diaphragm might fail as a force generator due to the slow regeneration of ATP.

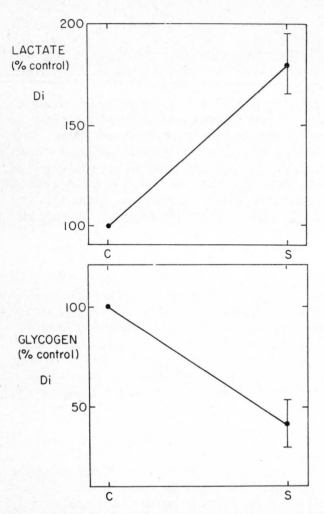

FIG. 4. Mean changes in the lactate and glycogen content of the diaphragm between the start of the experiment, before shock (control value, C), and the end of cardiogenic shock (S) in 12 dogs. Changes are expressed as % of control values. Bars, SEM.

Diaphragmatic fatigue in humans

Normal subjects breathing through a resistance develop diaphragmatic fatigue if the transdiaphragmatic pressure (Pdi) developed with each breath is greater than 40% of the maximum Pdi. Furthermore, when subjects are breathing 13% O_2, the endurance of the diaphragm decreases. If they are breathing at higher lung volume, the endurance of the respiratory muscles is again affected. In fact the critical pressure below which the respiratory muscles become fatigued at FRC is 50–70% of maximum pleural pressure, while at FRC plus half the inspiratory capacity, the critical pressure is 25–30% of the maximum (Roussos & Macklem 1977, Roussos et al 1979). The subjects in these studies were highly motivated, so we can assume that the neural drive remained adequate throughout the run and that the failure of the diaphragm was due to peripheral factors (peripheral fatigue). In an attempt to prove that the cause of fatigue was peripheral we stimulated one of the phrenic nerves transcutaneously before and after fatigue of the diaphragm. If central fatigue were responsible for the loss of force of the diaphragm, the Pdi generated by the same stimulus before and after fatigue would not be different. Fig. 5 shows that the Pdi during supramaximal stimulation (115 volts), with a

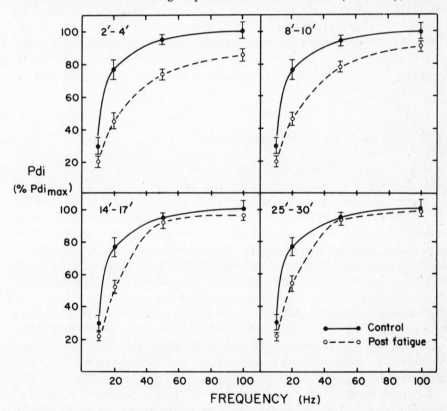

FIG. 5. Time course of the mean changes in the transdiaphragmatic pressure (Pdi)–frequency curves of the diaphragm after fatigue in four human subjects. The solid lines represent the averages obtained before fatigue (control). The broken lines represent the averages obtained at different times during the recovery period. Pdi is expressed as % of the Pdi generated for a stimulation frequency of 100Hz (Pdi_{max}). Bars, SEM.

pulse-width of 0.1 ms and 1–2 s duration at 10, 20, 50 and 100 Hz, was less after the diaphragm became fatigued. These results indicate that the loss of diaphragmatic strength at the end of the run is at least in part due to derangements at the periphery (at the neuromuscular junction or distal to it).

Edwards and his associates have hypothesized that the loss of force at a high frequency of stimulation (high frequency fatigue) is due to failure of the neuromuscular junction while the loss of force at a low frequency of stimulation (low frequency fatigue) is mainly due to failure of excitation–contraction coupling (Edwards 1978). The first type of fatigue is restored quickly, while low frequency fatigue persists for longer periods. We have tested the behaviour of the human diaphragm for one hour after the fatiguing run. Fig. 5 shows that the diaphragm responded in a similar manner as the other skeletal muscles: the Pdi at a low frequency of stimulation persisted for at least 30 min during the recovery period while the Pdi at 100 Hz recovered in 10–15 min. Although we cannot identify the mechanism underlying low and high frequency diaphragmatic fatigue from our experiments, our results support the hypothesis that diaphragmatic failure may at least in part be located in the periphery.

Electromyographic alterations during fatigue

During the last two decades frequency analysis has become an important tool in the evaluation of myoelectric activity. Frequency analysis has long been applied in detecting muscular fatigue. Piper in 1912 noted a decrease in the carrier frequency (the Piper rhythm) during a fatiguing contraction. Cobb & Forbes (1923) obtained the same results and also observed that the amplitude of the EMG increased. Kogi & Hakamada (1962) using frequency analysis demonstrated that this increase occurred in the low frequency region, while Kaiser & Petersen (1963) and Sato (1965) showed that the high frequency energy decreased during fatigue. These findings were carefully substantiated by Kadefors et al (1968). Recently, frequency analysis of the EMG has been used to detect fatigue of the diaphragm by Gross et al (1979, 1980). They found that the EMG power spectrum shifts when the diaphragm makes fatiguing contractions while it remains unchanged during contractions of low intensity.

We have applied frequency analysis to the EMGs recorded in the dogs described earlier. The shift in the EMG power spectrum was determined by taking the ratio of the power of a band with high (H) frequency energy (130–250 Hz) to the power of a band with low (L) frequency energy (20–40 Hz) (Fig. 6). As the power spectrum shifts to the left during fatiguing contractions the ratio decreases. Fig. 6 shows that the ratio falls very early in

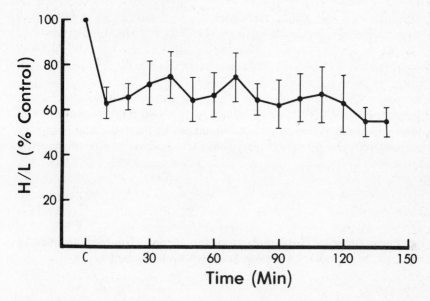

FIG. 6. Evolution in the ratio of high to low (H/L) frequency energy in the EMG power spectrum in 12 dogs during cardiogenic shock. Changes are expressed as % of the initial (control, C) value. Bars, SEM.

the experiment, before ventilation decreases or the diaphragm fails as a pressure generator. What are the mechanisms underlying the shift of the power spectrum during fatigue? There is no definite answer to this question. Lindström (1970) and Lindström et al (1970) claimed that the distribution of power between the different frequencies of the EMG power spectrum derived from the entire muscle bundle corresponds to the frequency distribution of power in the action potentials of the individual muscle fibres of which it is composed. Using the spectral 'dip' method, Lindström et al calculated that the power spectral shifts seen during fatigue could be accounted for by a progressive slowing of the conduction velocity, which would in turn prolong the action potential waveform of the fibre. Our recent work (Donovan et al 1979) does not support this theory of a reduction in conduction velocity. We have shown that cooling the muscles reduces the conduction velocity and in turn shifts the EMG to the left in the absence of fatigue. However, during fatigue, for the same changes in the power spectrum, there is little or no change in the conduction velocity. These results, coupled with the finding that the power spectrum shifts either with increasing load during apparently non-fatiguing contraction (Lindström et al 1970, Viitasalo & Komi 1977) or with the length of the muscle (Sato 1965), suggest that further studies need to be done to uncover the underlying mechanism of the power spectrum shift. Changes in the firing frequency or neuromuscular transmission (Edwards 1979), in conduction velocity (Lindström 1970, Lindström et al 1970, Mortimer et al 1970) and in synchronization (Person & Mishin 1964, O'Donnell et al 1973, Chaffin 1973) are some of the mechanisms that need to be studied.

In summary, the diaphragm fails as a force generator when the rate of energy demand exceeds the rate of energy supply. During respiratory failure in animals and humans diaphragmatic fatigue can be located in the periphery by means of phrenic stimulation or frequency analysis of the myoelectric signal.

Acknowledgements

This work was supported by grants from the Medical Research Council of Canada; C.R. is a scholar of the Medical Research Council.

REFERENCES

Aubier M, Trippenbach T, Sillye G, Viires N, Roussos CS 1980 Respiratory failure in cardiogenic shock. Proc Int Union Physiol Sci 14:305 (abstr)

Bradley ME, Leith DE 1978 Ventilatory muscle training and the oxygen cost of sustained hyperpnea. J Appl Physiol 45:885-892

Chaffin DB 1973 Localized muscle fatigue—definition and measurement. J Occup Med 15:346-354

Cobb S, Forbes A 1923 Electromyographic studies of muscular fatigue in man. Am J Physiol 65:1207-1222

Donovan E, Roussos Ch, Andersen J, Macklem PT, Ritchie B 1979 The relation between EMG power spectrum and conduction velocity during human muscle fatigue. Fed Proc 38:1382 (abstr)

Edwards RHT 1978 Physiological analysis of skeletal muscle weakness and fatigue. Clin Sci Mol Med 54:463-470

Edwards RHT 1979 The diaphragm as a muscle: mechanism underlying fatigue. Am Rev Respir Dis 119:81-84

Fitts RH, Holloszy JO 1976 Lactate and contractile force in frog muscle: development of fatigue and recovery. Am J Physiol 231:430-433

Fuchs F, Reddy V, Briggs FN 1970 The interaction of cations with the calcium-binding site of troponin. Biochim Biophys Acta 221:407-409

Gross D, Grassino A, Ross WRD, Macklem PT 1979 Electromyogram pattern of diaphragmatic fatigue. J Appl Physiol 46:1-7

Gross D, Ladd H, Riley E, Macklem PT, Grassino A 1980 Effect of training on strength and endurance of the diaphragm in quadriplegics. Am J Med 68:27-34

Kadefors R, Kaiser E, Petersen I 1968 Dynamic spectrum analysis of myo-potentials with special reference to muscle fatigue. Electromyography 8:39-74

Kaiser E, Petersen I 1963 Frequency analysis of muscle action potentials during tetanic contraction. Electromyography 3:5-17

Kogi K, Hakamada T 1962 Slowing of surface electromyogram and muscle strength in muscle fatigue. Report of the Institute of Science of Labour (Kurashiki) 60:27-41

Lindström L 1970 On the frequency spectrum of EMG signals. Thesis, Research Laboratory of Medical Electronics, Göteborg

Lindström L, Magnusson R, Petersen I 1970 Muscular fatigue and action potential conduction velocity changes studied with frequency analysis of EMG signals. Electromyography 10:341-355

Mortimer JT, Magnusson R, Peterson I 1970 Isometric contraction, muscle blood flow, and the frequency spectrum of the electromyogram. In: Spring E, Jauhiainin J (eds) Proc 1st Nordic Meeting on Medical and Biological Engineering. Finnish Society for Medical and Biological Engineering, Helsinki, p 142-144

Nakamura Y, Schwartz S 1972 The influence of hydrogen ion concentration on calcium binding and release by skeletal muscle sarcoplasmic reticulum. J Gen Physiol 59:22-32

O'Donnell RD, Rapp R, Berkhout J, Adey WR 1973 Autospectral and coherence patterns from two locations in the contracting biceps. Electromyogr Clin Neurophysiol 13:259-269

Person RS, Mishin LN 1964 Auto- and cross-correlation analysis of the electrical activity of muscles. Med Electron Biol Eng 2:155-159

Piper H 1912 Electrophysiologie menschlicher Muskelin. Verlag von Julius Springer, Berlin, p 126

Roussos CS, Macklem PT 1977 Diaphragmatic fatigue in man. J Appl Physiol 43:189-197

Roussos CS, Fixley M, Gross D, Macklem PT 1979 Fatigue of inspiratory muscles and their synergic behavior. J Appl Physiol 46:897-904

Rudolph AM, Heymann MA 1967 Circulation of the fetus in utero: methods for studying distribution of blood flow, cardiac output and organ blood flow. Circ Res 21:163-184

Sato M 1965 Some problems in the quantitative evaluation of muscle fatigue by frequency analysis of the electromyogram. J Anthropol Soc Nippon (Reprinted in English) 73:20-27

Trivedi B, Danforth WH 1966 Effect of pH on the kinetics of frog muscle phosphofructokinase. J Biol Chem 241:4110-4112

Viitasalo JHT, Komi PV 1977 Signal characteristics of EMG during fatigue. Eur J Appl Physiol Occup Physiol 37:111-121

Whitelaw WA, Derenne JPh, Milic-Emili J 1975 Occlusion pressure as a measure of respiratory center output in conscious man. Respir Physiol 23:181-199

DISCUSSION

Jones: Since you measure the neurogram from the phrenic nerve, can you also do a power spectrum analysis, and does it change with fatigue in the same way as the muscle power spectrum? It would be interesting to know this, as one of the possible explanations of the shift in the muscle power spectrum is that it is due to a change in the frequency of activation of the muscle.

Roussos: We tried to do this in several dogs, using this model. We have great difficulties in obtaining pure phrenic nerve activity. At any rate, whatever the signal was, we haven't been able to see any change in the power spectrum.

Saltin: You mentioned the glycogen content of fibres in the diaphragm. Do you know whether some fibres are completely empty and others have glycogen left?

Roussos: We are working on this now. We have no information yet on which fibres are depleted.

Saltin: It could be that a substantial proportion of the fibres are completely depleted of glycogen in this animal model (cardiogenic shock and respiratory failure).

Dawson: How does aminophylline increase the force in the diaphragm?

Roussos: If the loss of force is due to some extent to a failure of excitation–contraction coupling—in other words, to calcium sequestration—perhaps aminophylline releases calcium and thus potentiates the contraction.

Jones: Aminophylline is a preparation of theophylline. Its action in potentiating the muscle twitch is almost identical to that of caffeine.

Dawson: You have the problem, then, that methylxanthines also affect the adenylate cyclase system.

Jones: The action of caffeine and other methylxanthines in potentiating the muscle twitch and increasing low frequency force is almost certainly not mediated by cyclic AMP. No correlation was found between ability to inhibit phosphodiesterase activity and to potentiate the twitch (D. A. Jones & S. Howell, unpublished observations).

Dawson: Is that true of aminophylline as well?

Roussos: Yes. All xanthine compounds have been tested, and an effect through AMP does not appear to be involved in their mechanism of action (Kramer & Wells 1980).

Edwards: We have only tangentially mentioned neuromuscular transmission failure so far. We have heard mostly about factors beyond the neuromuscular junction that play a role in high frequency fatigue. Perhaps John Stephens would summarize his evidence for failure of neuromuscular transmission as a cause of fatigue?

Stephens: In our study of the mechanism of fatigue during maintained maximum voluntary isometric contraction (MVC) of the first dorsal interosseous muscle of the hand, Taylor and I made the following observations (Stephens & Taylor 1972, 1973a, b). Fatigue occurs in two phases. In the first, lasting about a minute, force falls to about 50%. Natural EMG activity falls with the same time course and the normal linear relation between force and electrical activity is preserved. In the second phase, force falls relatively faster than muscle electrical activity.

Two simple mechanisms can be put forward to account for the fact that during the first stage of fatigue, EMG activity falls linearly with the same slope with respect to force as in unfatigued muscle. First, there might be some reduction in motor output from the central nervous system, despite the subject's best efforts. Secondly, there might be failure of neuromuscular transmission. Neither of these mechanisms would be expected to result in a change in the relationship between force and EMG as recorded in the unfatigued muscle.

In order to distinguish between these possibilities we recorded the electrical response of the muscle to maximal nerve stimulation. Surface-recorded muscle action potentials evoked by ulnar nerve stimulation were reduced on average to about 65% of normal, most of the reduction occurring during the first minute of a maintained MVC.

These findings led us to conclude that during the first stage of fatigue in an MVC there is neuromuscular transmission failure. In the second phase of fatigue, where natural EMG activity falls less rapidly than force, we concluded that failure of the contractile element becomes progressively more important.

Since these observations were made, Dr Grimby has recorded single motor unit activity during a maintained MVC (this volume, p 157-167) and shown quite clearly that motor output is reduced in these circumstances, confirming the observation made by Marsden et al (1971) that motor unit firing rate falls progressively at the start of a fatiguing contraction. Close inspection of these recordings shows not only a progressive reduction in the mean interval between spikes but also occasional long intervals where an action potential appears to be missing. I would interpret these gaps in the recorded spike train as being due to failure of transmission of the motor impulse to the muscle fibre whose electrical activity contributed the major part of the recorded action potential.

In conclusion, then, I think there is now evidence for a reduction in motor output and perhaps some direct evidence for neuromuscular transmission failure during the first stage of fatigue during sustained maximum voluntary effort. The question remains of the relative quantitative importance of these two mechanisms in determining the parallel loss of force and electrical activity which is observed in this type of contraction.

Bigland-Ritchie: In your published paper (Stephens & Taylor 1972) the mass action potentials recorded after 45 seconds of maximal contraction are as large in total area as the prefatigue controls (+80% MVC) at a time when the force was said to have declined by about 40%. They are also larger than those that precede or follow them. Were you sure your stimulus was always maximal? Dr Lippold and I have had problems with this. It is easy to stimulate maximally when the hand is relaxed; but during maximal contractions the tendons over the ulnar nerve pull up, the wrist tends to bend and the electrodes get displaced. We frequently recorded M waves that declined in size; but we were always able to bring them back by repositioning the electrodes and turning up the stimulus intensity during the contraction. This possibility must be ruled out before one can prove that M wave decline is due to neuromuscular block. We have now also recorded M waves intramuscularly (unpublished results). These cannot have originated from contamination by other, non-fatigued muscles.

Secondly, not all Dr Stephens' plots of the decline in naturally occurring EMG are equally smooth. Some showed transient increases up to the original maximal level after the force had fallen by 60–70% (Stephens & Taylor 1972, Fig. 5b and d), or were above control values during the first 30% of force loss even with the circulation occluded (Fig. 5c). We did studies (Bigland-Ritchie et al 1975) in which brief 'super-efforts' during the course of fatigue momentarily reversed the normal decline in EMG. This decline was not due to a lack of voluntary effort either in our experiments or in Dr Merton's (1954), but can be briefly overcome during ballistic efforts. These observations are difficult to explain if the loss of force is due to neuromuscular block.

Stephens: A number of authors have described a progressive reduction in motor unit EMG amplitude during a period of repetitive stimulation. The frequency and duration of stimulation at which the decline in EMG occurs varies for different motor units (Wuerker et al 1965, Stephens et al 1973, H. P. Clamann, this volume, p 148). Such reductions in unit EMG amplitude can be observed at quite modest stimulus frequencies, say 40 p.p.s. Similar changes in motor unit EMG amplitude can be expected to be taking place during the maintained maximum voluntary contraction in man. Whatever the cause of reduction in unit EMG amplitude, it can be expected to lead to a reduction in the nerve-evoked muscle mass action potential. The available evidence at the motor unit level is in accord with the observation that the nerve-evoked mass action potential is reduced during maintained maximum voluntary contraction and at odds with Dr Bigland-Ritchie's observation that there is no change. We won't find out how much the reduction in EMG is due to transmission failure until we make recordings specifically to test neuromuscular transmission. We should look at individual motor units, remembering that the neuromuscular junctions are different for the FR, FF and S unit types. We should be looking for differences in unit susceptibility to transmission failure and relating this to recruitment order and motor unit mechanical properties. We should be doing the sort of experiment that Dr Clamann is doing (p 148) and studying transmission failure directly at the single unit level.

Edwards: You haven't mentioned 'jitter' yet! (Jitter is the variability in the time interval between the action potentials from two muscle fibres belonging to the same motor unit.) An attempt to get an independent assessment of impaired neuromuscular transmission has been based on jitter, which appears to be increased with ischaemia (Dahlbäck et al 1970), in circumstances in which we are demonstrating force fatigue. To that extent there may be a contribution from impaired neuromuscular transmission but, for the reasons given by Brenda Ritchie, this is unlikely to be the only cause.

Stephens: The crucial question is how important quantitatively are the reductions in firing frequency seen in unit studies in accounting for the reduction in force in the first minute of a fatiguing contraction, and how important are neuromuscular transmission failure and failure of the contractile machinery. We shall waste time and effort if we persist in looking for a single site of fatigue. It will be different in different types of muscle fibre and in different sorts of fatiguing contraction.

Edwards: I agree. But are there limitations to the single-shock approach to assessing failed neuromuscular transmission? We have used the programmed stimulation myogram to separate low and high frequency fatigue, and as a method it stresses neuromuscular transmission to an extent that might not be evident with a single shock. Have you looked at the effects of tetani?

Stephens: No. The major limitation of this sort of experiment is that the action potentials cannot be interpreted simply. Our interpretation in terms of neuromuscular transmission failure is receiving support from direct observation. It is not the whole story, however.

Edwards: There is also conflicting information from direct stimulation of human muscle (Hill et al 1980), since these experiments show clear evidence of changes beyond the neuromuscular junction.

Stephens: Certainly, and the site of fatigue depends crucially on the details of the contraction as it is being made.

Bigland-Ritchie: I agree. However, may I re-emphasize the problem of not knowing precisely what tetanic frequencies are appropriate for testing after different types of fatigue? Dr Stephens refers to 40/s as being 'modest'. After fatigue we did not observe natural firing rates above about 10–15/s, where block is unlikely.

Grimby: In our single-unit experiments we have two groups of subjects. One group of subjects had abnormal motor units and were trained to maintain maximal tension voluntarily. In these subjects we saw loss of components of the motor unit potentials, suggesting transmission blockade during maximum voluntary effort. But this may be due to the abnormality of the motor units. The other subjects had normal motor units but were not trained to maintain a strong voluntary effort. In these subjects we saw no signs of transmission blockade, but this may be due to the lower level of voluntary drive.

Edwards: Can we go on to the interpretation of the shift of the power spectrum of the EMG towards lower frequencies in a sustained MVC of, say, the quadriceps muscle? We know that the firing frequency is being reduced because of central mechanisms. According to Brenda Ritchie there has been no change in the conduction velocity of the action potential, or much less than with stimulation. The argument originally made for the high/low ratio being a function of lactate accumulation (Mortimer et al 1970), which is supposed to alter conduction velocity, is thus not tenable. Dr Wiles will also show this later (p 264-277) with evidence from patients with McArdle's disease (myophosphorylase deficiency) who cannot make lactic acid. We then worked with Peter Macklem and Charis Roussos to see what other factors are involved. The question is to what extent the central firing frequency is a determinant of the alteration in the ratio.

Roussos: If the firing frequency changes, the power spectrum of the phrenic neurogram will probably change. In a limited number of experiments, as I said, we see no indication of a shift in the power spectrum.

Edwards: The frequency of muscle action potentials depends on how effectively the sarcolemma is excited. When there is high frequency fatigue (i.e. impaired excitation at high frequencies), there appears to be a shift in the power spectrum, according to what you have shown us.

Roussos: Yes. We have also done experiments in which we made MVCs. During the 30 seconds when there is no indication of low frequency fatigue, and only high frequency fatigue, we have the greatest decrease in the high:low ratio.

Edwards: Another point of some interest: Dr David Pengelly (personal communication) told me that in partial curarization in man there is a shift in the high:low ratio in the direction of low frequencies. So in fresh muscle, therefore, when there is no question of alterations in conduction velocity or lactate accumulation, there is a similar shift to that in fatigue. I have argued that we may be seeing a low-frequency band-pass filter effect at the time when a muscle is showing high frequency fatigue. In other words, the filter allows only low frequencies to appear and not high ones.

Moxham: I didn't think that the experiments we did on the effects of partial curarization with Dr Roussos in Montreal supported that interpretation.

Roussos: In some contractions we saw the change in the power spectrum and in some we didn't. The contractions where we showed the power spectrum shift were all low force contractions in absolute terms, but for that degree of curarization they were almost maximal. So the absolute force was low but for the working motor units it was very high.

Moxham: That would be consistent with a turning down of central firing frequency. Another observation is that if you produce low frequency fatigue in a muscle, so that subsequently more drive is required to produce a given force, and if you look at the high:low ratio in the recovery period, it is usually increased for submaximal forces.

Roussos: This fits with the idea that if the firing frequency increases but the muscle is not fatiguing, you have more energy in the high frequency area and so the high:low ratio will increase.

Edwards: The point about this ratio is that it is a tool which is easy to use technically and so could be applied to studies of fatigue in patients. It is important to try to agree about what it means in physiological terms. We have heard that this change takes place before the force has fallen, and to that extent it could be a useful guide to fatigue. One question is to what extent it reflects alterations in the appearance (or disappearance) of the action potential, and can indicate recovery of vital processes. We know that the action potential recovers quickly.

Lippold: Can Dr Roussos please say what the high:low ratio really means?

Roussos: It is an index for quantifying the shift in power spectrum, being the ratio of the power in the spectrum in a high and a low frequency band.

Lippold: Can you say how the high:low ratio differentiates between the repetition frequency of discrete events (i.e. the firing frequency of units) and the shape of the waveform of the event (i.e. the rise time of single unit spikes)? Or is the result an amalgam of the two?

Edwards: It is an empirical observation which may give an indication or a warning of fatigue. In the absence of simultaneous force and electrical measurements, it is not possible to consider the site of fatigue.

Stephens: People have been studying the power spectrum of the EMG for 20 years, but there has never yet been a satisfactory explanation of why it is the shape it is, and why it changes in the way it does in various circumstances. The reason is that it is an extremely complicated function made up of many different additive factors. In my view, to seek a physiological explanation for it is a lost cause.

If you want to understand a physiological process, you make a measurement the basis of which you understand. If you are interested in the relationship between force and metabolism, you measure force and you measure metabolism. You try to measure two quantities, both of which you understand. An EMG power spectrum is something that nobody understands! It is not a good measure of anything, and I don't think it ever will be, because it's too complicated.

Roussos: So we mustn't even try to understand it?

Macklem: It all depends what you want to use it for. If one is faced with the problem of diagnosing inspiratory muscle fatigue in a patient with acute respiratory disease, a non-invasive technique which showed changes by which you can either predict or diagnose fatigue would be exceedingly useful, even though empirical. My own feeling, in fact, is that this will probably not be useful diagnostically. Clinical signs, such as that mentioned by John Moxham (abdominal paradoxical motion and an alternation between taking breaths with the rib cage and the abdomen), will probably be more powerful in diagnosing inspiratory muscle fatigue than the EMG.

Merton: If you have a repeated waveform and take the power spectrum of that, the spectrum will contain components related to the frequencies present in the basic waveform, together with components related to the repeat frequency, and composite terms. The difficulty in interpreting the power spectrum from muscle is that the basic waveform, the muscle action potential, lengthens out during fatigue at the same time as the repeat frequency is falling. It is not out of the question to disentangle these factors, but it has not yet been done.

Edwards: In Montreal we tried to observe the power spectrum during different frequencies of stimulation, to see how much spectral shift would occur when fatigue was produced with constant frequency stimulation. Unfortunately everything was dominated by the frequency of stimulation.

Merton: All that the high:low ratio reflects is that if you listen through a loudspeaker to the electromyogram of a fatiguing muscle, you can hear the predominant pitch of the note drop as the muscle fatigues. The ratio is simply an index of that drop in pitch.

Bigland-Ritchie: In our experiments (Donovan et al 1979) we changed the conduction velocity and thus the waveform of the action potential by cooling instead of by fatigue. We found that when the same slowing seen in fatigue was induced instead by cooling, there was a much smaller shift in the power spectrum. Thus factors other than action potential waveform alone must also be involved. As Dr Stephens has said, the process in fatigue is more complex than is often supposed (Lindström et al 1977, Broman 1977).

Stephens: If you want to study slowing of conduction along muscle fibres, you should record it directly. If you want to study neuromuscular transmission failure, study it directly. Don't try to use synchronous muscle action potentials, because the arguments start to get full of holes. We now have the techniques and background knowledge to be able to do specific experiments; we don't have to use these indirect methods.

Moxham: We should add, in fairness to Peter Macklem's point, that Dr Stephens is talking as a pure physiologist, and the fact remains that for the clinician looking after patients with respiratory failure a non-invasive technique for diagnosing respiratory muscle fatigue would be very useful.

REFERENCES

Bigland-Ritchie B, Hosking GP, Jones DA 1975 The site of fatigue in sustained maximal contractions of the quadriceps muscle. J Physiol (Lond) 250:45P-46P

Broman H 1977 An investigation on the influence of a sustained contraction on the succession of action potentials from a single motor unit. Electromyogr Clin Neurophysiol 17:341-358

Dahlbäck L-O, Ekstedt J, Stålberg E 1970 Ischemic effects on impulse transmission to muscle fibers in man. Electroenceph Clin Neurophysiol 29:579-591

Donovan E, Roussos Ch, Anderson J, Macklem PT, Ritchie B 1979 The relationship between EMG power spectrum and conduction velocity during human muscle fatigue. Fed Proc 38:1382 (abstr)

Hill DK, McDonnell MJ, Merton PA 1980 Direct stimulation of the adductor pollicis in man. J Physiol (Lond) 300:2P-3P

Kramer GL, Wells JN 1980 Xanthines and skeletal muscle: lack of relationship between phosphodiesterase inhibition and increased twitch tension in rat diaphragms. Mol Pharmacol 17:73-78

Lindström L, Kadefors R, Petersén I 1977 An electromyographic index of localized muscle fatigue. J Appl Physiol 43:750-754

Marsden CD, Meadows JC, Merton PA 1971 Isolated single motor units in human muscle and their rate of discharge during maximal voluntary effort. J Physiol (Lond) 217:12P-13P

Merton PA 1954 Voluntary strength and fatigue. J Physiol (Lond) 128:553-564

Mortimer JT, Magnusson R, Petersén I 1970 Conduction velocity in ischemic muscle: effect on EMG frequency spectrum. Am J Physiol 219:1324-1329

Stephens JA, Taylor A 1972 Fatigue of maintained voluntary muscle contraction in man. J Physiol (Lond) 220:1-18

Stephens JA, Taylor A 1973a The relationship between integrated electrical activity and force in normal and fatiguing human voluntary muscle contractions. In: Desmedt JE (ed) New developments in electromyography and clinical neurophysiology. Karger, Basel, vol 1:623-627

Stephens JA, Taylor A 1973b Analysis of muscle fatigue mechanisms in man. In: Somjen GG (ed) Neurophysiology studied in man. Excerpta Medica, Amsterdam, p 412-419

Stephens JA, Gerlach RL, Reinking RM, Stuart DG 1973 Fatiguability of medial gastrocnemius motor units in the cat. In: Stein RB et al (eds) Control of posture and locomotion. Plenum Press, New York, p 179-185

Wuerker RB, McPhedron AM, Henneman E 1965 Properties of motor units in a heterogeneous pale muscle (m. gastrocnemius) of the cat. J Neurophysiol 28:85-99

The tremor in fatigue

OLOF LIPPOLD

Department of Physiology, University College, London WC1E 6BT, UK

Abstract After a maximal voluntary effort made for about 2 min, tremor of the muscles concerned is increased in amplitude by up to one order of magnitude for a period of several hours afterwards. Spectral analysis reveals that all frequencies of tremor show this increase. Maximal electrical stimulation of the motor nerve to the muscle does not result in any change in tremor so it is inferred that the increase is due to the operation of spinal or supra-spinal mechanisms.

A sub-maximal voluntary effort, maintained for about 1 h, leads to the development of large amplitude, low frequency tremor (4–6 Hz) in addition to increased physiological tremor (8–12 Hz). It is generally accepted that physiological tremor originates as an oscillation in the reflex arc servo loop. The 4–6 Hz slow tremor of this type of fatigue also appears to arise as servo loop oscillation, the feedback delays being longer than in physiological tremor. The spectrum of the slow tremor resembles that found in Parkinsonism and it may be a useful model for the tremor of that condition.

Introduction

The near-maximal usage, or the prolonged usage, of a muscle leads to several changes in that muscle's function, collectively known as fatigue. Most of these changes are in the form of functional muscular deficits of short or long duration, but since the central nervous system is also involved in this kind of activity of muscle, changes referable to motor control mechanisms are also part of the picture seen in fatigue. In this paper are described the effects that (1) strong and (2) prolonged muscle contraction bring about upon tremor.

The definition and measurement of tremor

Tremor can be defined as the error superimposed on the movement of a limb having a frequency above 1/second. It is caused by:

1981 Human muscle fatigue: physiological mechanisms. Pitman Medical, London (Ciba Foundation symposium 82) p 234-248

234

(a) oscillations in the force of muscular contraction;

(b) the mechanical excitation of resonances in the limb due to inertia and elasticity of muscles, tendons and joints;

(c) the transmitted cardiac pulse.

The proportion of tremor due to each of these mechanisms varies with the way in which the tremor is recorded and also, to some extent, with which limb is involved.

In order to measure normal muscular tremor (a) it is necessary to ensure that the limb concerned is held in isometric conditions. This eliminates any movements due to mechanisms (b) and (c). Commonly, a strain gauge is used to measure tremor force at specified levels of voluntary excitation. For measurement of tremor of mechanical origin (b) and cardiac origin (c) an accelerometer is normally used, fixed to the freely moving limb. Normal muscular tremor, which has a frequency of 8–12 Hz, cannot usually be observed in a limb whose inertia is so large that it is unable to oscillate at frequencies as high as this. For example, the forearm or leg will not have tremor at frequencies much above 3–5 Hz. A confounding instance of this effect occurs at the wrist. The relaxed wrist will, in most people, oscillate at about 9 Hz when a force is applied to it, as a result of its stiffness and inertia. Since this is about the same frequency as normal muscular tremor (8–12 Hz), the two mechanisms reinforce each other.

In experiments quoted in this paper, isometric tremor force was measured for extension of the finger, or the electromyogram (EMG) from the finger extensor muscles was analysed, in both cases while the voluntary strength was kept at a constant, low level.

The effect of brief, strong contraction on tremor

Most people have experienced the fact that heavy muscular work results in a varying degree of increased tremor in the muscle groups involved. Often the increase lasts for hours or even days after a comparatively brief contraction, if this is near the maximum force level. For example, those engaged in delicate manual tasks, such as neurophysiologists dissecting single fibres, often notice that their performance is impaired by the tremor produced by playing squash or digging the garden on the previous day. Indeed, this phenomenon of increased tremor after strong contractions was studied by Bousfield as long ago as 1932.

With all the known deficits occurring in a muscle when it is fatigued, one might predict that the tremor increase would be due to actions within the muscle itself. Experiments show, however, that this is not so, but that the origin appears to be in the central nervous system. The detailed analysis of the action of fatigue on tremor can be found in Furness et al (1977).

Frequency analysis of the tremor of effort

When a weight of about 1 kg was lifted with the finger for 2 min the long-lasting increases in tremor that resulted were obvious, even without analysis. In some of the subjects the effect could still be observed 4 or 5 h later.

The degree of increase in tremor appeared to depend upon the amount of 'effort' expended rather than the actual performance of the experimental subject. Many subjects were unable to maintain a contraction strength of 1 kg for 2 min; these subjects were exhorted to continue making a maximal effort for the 2 min even though the weight had been allowed to drop.

Frequency analysis of the signal from a strain gauge (Lippold 1970) showed that all frequencies of tremor of the finger were enhanced in their amplitude, in most cases by about an order of magnitude.

Fig. 1 shows the analysis of tremor after 2 min of near-maximal effort in a

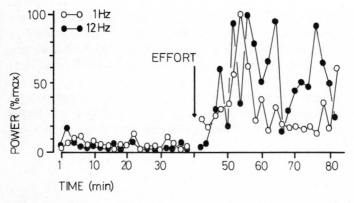

FIG. 1. The time course of the increase in tremor after a brief, maximal voluntary 'effort' of m. extensor digitorum communis. The finger is raised with a force of 0.5 N against a strain gauge for the purpose of tremor measurement every 2 min before and after the effort. Open circles are power of tremor force at 1 Hz; filled circles at 12 Hz (which was the value of the peak frequency in the 8–12 Hz band in this subject). Power is expressed as % maximum power at each of the frequencies.

Three hours later the powers had still not returned to control levels.

(From experiments by Furness et al 1977.)

subject whose normal muscular tremor showed a marked peak at about 12 Hz. From 0 to 40 min are shown control values of the amplitudes of tremor up to 1 Hz (open circles) and at 12 Hz (closed circles). At 40 min, 2 min of effort were made. It can be seen that for the subsequent 40 min there is roughly a tenfold increase in amplitude of the components at the two frequencies.

It might be thought that increased tremor produced by a brief, strong effort would involve a generalized nervous component or possibly be due to some circulating hormonal effect, for example, of adrenaline, which is known to increase tremor. However, recording from the contralateral finger shows that this is not the case, for tremor there is not increased.

Tremor produced by motor nerve stimulation

On the supposition that the effects of a strong effort as described above originate peripherally in the muscle, or in the neuromuscular junction, it would be expected that a tetanic electrical stimulation of comparable strength and duration would also lead to a period of increased tremor.

It is possible to stimulate the motor nerve to the muscle with $50 \mu s$ pulses at 30 Hz and a voltage large enough to cause the finger to raise the 1 kg weight. When this was done for 2 min, frequency analysis showed that tremor was never increased by this procedure; indeed there was some evidence that a decrease often follows.

Tremor after 'effort' without contraction

It has already been mentioned that it is the 'effort' made by the subject that is of importance in the generation of long-term after-effects on tremor amplitude. One can test this in the absence of muscle activation by first blocking conduction in the motor nerve to the muscle and then instructing the subject to make as strong an effort as possible to contract the muscle even though it then produces no movement. Block can easily be induced by inflating a cuff on the upper arm to above systolic arterial pressure for about 45 min. On its own, this procedure does not have any subsequent effect upon tremor amplitude.

Provided he cannot see his finger, the subject usually thinks that it has actually moved during the exertion of the effort. Subsequent frequency analysis then shows an increase in tremor of much the same time course and magnitude as was found previously in weight-lifting by the finger. It might be objected that the stress involved in being subjected to an inflated arterial cuff for 40 min would be enough to induce an increase in tremor. However, this is not the cause of the increase, because when cuffs are applied to both arms but the effort is made on one side, it is only in the finger on that side that an after-effect on tremor is found.

Synchronization of motor unit firing with fatigue

The experiments described so far can only be interpreted as demonstrating that the increase in tremor following a brief, strong effort has its origin in central nervous mechanisms and that it cannot be due to any changes within the muscle itself, or in the neuromuscular junction. What mechanisms might be invoked to explain such an increase in tremor? Since all frequencies are involved, it cannot be the consequence of any alterations in the properties of the reflex arc alone, or of the fusimotor system, because these would have effects confined to the 8–12 Hz frequency band.

An increase in tremor associated with a given voluntary contraction strength indicates that there is an increased degree of synchronization between the firing of motor units. It might also suggest that after the brief effort, different motor unit types are active during the tremor recordings. If larger units are recruited, increased tremor would be expected because of the extra force that each unit develops.

Examination of the electrical activity from surface electrodes over the muscle shows that there is a change in its nature during and after a strong voluntary contraction. During an intense effort large 'waves' appear and reach a maximum amplitude after about 2 min; these are the so-called W-waves of Jessop & Lippold (1977).

Fig. 2 shows a recording of the surface electrical activity from m. abductor digiti minimi brevis taken about 1.5 min after the commencement of a maximal voluntary isometric contraction. The prominent W-waves can be seen, separated by periods of up to 50 ms of electrical silence. Also shown in Fig. 2 is an M-wave, which is the maximal electrical response to supra-maximal stimulation of the motor nerve (ulnar nerve at the elbow). It can be seen that W-waves may be of about the same amplitude as the M-wave. Their duration is also similar, but W-waves are always slightly longer. This indicates that during a maximal effort, up to 100% of the motor units are being activated simultaneously.

Electrical activity recorded from the muscle during a test contraction of 2 N made before the effort shows the normal irregular EMG pattern; afterwards it shows a pattern of larger waves, or groups of waves, interspersed with silent periods. This change in appearance shows that the degree of synchronization has altered, and it can be inferred that the long-term increase in tremor resulting from a brief, strong effort is, at least partly, due to some degree of persistence of the increased synchronization of motor units observed to occur actually during the effort.

There is, as yet, no information on the locus or the mechanism of this change in synchronization.

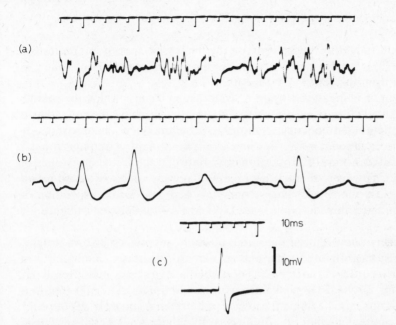

FIG. 2. Synchronization of motor unit firing. The three traces (a), (b) and (c) are records of the electrical activity recorded from surface electrodes over the belly of m. abductor digiti minimi brevis. Record (a) is a control taken at the beginning of a maximal voluntary isometric contraction against a strain gauge. Record (b) is taken 1.5 min later under the same conditions. Note silent periods between large W-waves, indicating grouping of motor unit action potentials, i.e. greatly increased degree of synchronized firing of motor units. Record (c) is an M-wave (supra-maximal stimulation of the ulnar nerve at the elbow), taken at 1.5 min after the start. The amplitude and duration of the W-waves are of about the same size as those of the M-wave. This must mean that about 100% of motor units are firing at almost the same instant to produce a W-wave. See Jessop & Lippold (1977) for details.

The effect of prolonged contraction on tremor

In 1897 Eshner first showed that large tremor has a lower frequency than small amplitude tremor. If the wrist is held out beyond the edge of a table supporting the forearm, after about an hour a large tremor develops, as described by Stiles (1976). The latter author has analysed this fatigue tremor in detail and has shown that it has a predominant frequency around 4–6 Hz. Unfortunately, the use of tremor of the hand (rotating about the wrist joint) tends to render analysis difficult; for reasons already mentioned, the mechanical and reflex components are confounded.

Frequency analysis of prolonged contraction tremor

In order to study the mechanism of the increased tremor of prolonged contraction it is necessary to use finger tremor in order to separate the mechanical from the reflex components.

When tremor is measured using a strain gauge against which the middle finger is extended with a force of 0.5 N, frequency analysis shows that a declining amplitude–frequency relationship is obtained that is approximately fitted by the equation $a = k.1/f^2$, where a is amplitude, f is frequency and k is an arbitrary constant of proportionality. Between the frequency values of 8 and 12 Hz, however, there is an excess of power, which is the so-called 'physiological tremor peak' and is commonly accepted as the expression of oscillation in the reflex servo loop (see Lippold 1973 for detailed evidence for this).

If the finger is now kept in the raised position, supporting its own weight, with the wrist fixed rigidly so that only m. extensor digitorum communis is in use, tremor recording at intervals as described above shows that after about 10 to 20 min, the 8–12 Hz peak increases in amplitude. After 30–90 min a second peak appears, having a frequency of between 4 and 6 Hz. Both peaks are present at the same time. It is quite clear that there is not a gradual change in frequency of the peak initially at 8–12 Hz. From its first appearance to the termination of the experiment, the frequency of the second peak (the 4–6 Hz one) remains at approximately the same value in any given experimental subject. The frequency of the first peak (the 8–12 Hz one) also does not change throughout the experiment.

These conclusions can also be demonstrated for isotonic recording from the finger, a convenient method for this being to rectify and demodulate the EMG potentials from the long extensor muscles of the finger. Fig. 3 shows the frequency analysis obtained in this way at the commencement of finger raising and 15 and 45 min later.

The origin of the 4–6 Hz tremor of prolonged fatigue

The 4–6 Hz tremor following a prolonged sub-maximal contraction could arise as either (a) a visco-elastic mass mechanism or (b) a neural or a feedback mechanism involving the stretch reflex (and possibly skin and joint receptors).

Alternative (a) is unlikely to be an important factor and cannot be responsible for any of the 4–6 Hz tremor shown in Fig. 3 because the recording here is solely of bursts of electrical activity from the extensor muscle.

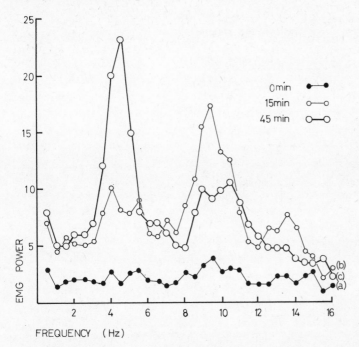

FIG. 3. Frequency analysis of tremor of m. ext. dig. comm. at the beginning (a), after 15 min (b), and after 45 min (c), of a submaximal contraction strong enough to raise the middle finger. The three graphs are the powers of the EMG amplitude (proportional to tremor force; see Lippold 1973) after rectification and demodulation (using a third-order low pass Butterworth filter with the roll-off at 20 Hz and −3dB/octave). Values were subject to digital smoothing by thirds.

Note that with this analysis the 'equal power curve' is a straight line parallel to the X-axis and not a function of $1/f^2$ as is the case for an 'equal force curve' of tremor force analysis.

The graphs show (a) very little tremor, (b) a considerable increase in tremor at 11 Hz and (c) two tremor peaks, one at 4 Hz and one at 11 Hz. Since these two peaks coexist in curve (c), there can be no question of a gradual change in frequency from 11 Hz to 4 Hz as fatigue progresses.

Alternative (b), a neural mechanism, could either arise centrally in the form of a pacemaker, say in the motor cortex or basal ganglia, or it could represent the operation of an unstable servo loop. These two possibilities can be distinguished by applying a step-function perturbation to the system and observing whether the responses are in phase with this input (Lippold 1970). Fig. 4 shows the result of a sudden stretch applied to the extended finger (a solenoid is used to give rapid flexion at the metacarpophalangeal joint). The mechanical input was repeated 32 times every 1913 ms and the resultant oscillatory response was averaged with respect to the instant of stretching for a duration of about 1.2 s.

The control recording shows relatively little summated activity at any

(a)
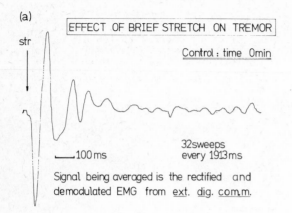

str

EFFECT OF BRIEF STRETCH ON TREMOR

Control : time 0min

└──100 ms

32sweeps
every 1913 ms

Signal being averaged is the rectified and
demodulated EMG from ext. dig. com.m.

(b)
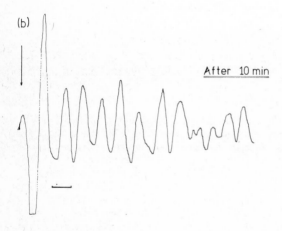

After 10 min

(c)
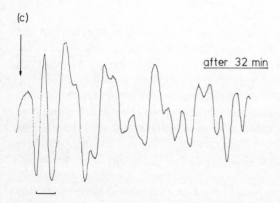

after 32 min

frequency, but a damped train of three or four waves can be observed at approximately 10 or 11 Hz. After 15 min the whole sweep is occupied by large waves at 10–11 Hz which have truly summated; that is, they are in phase with each other and with the input. After 45 min large waves at about 4 Hz have appeared in the average (in addition to the 10–11 Hz waves which are simultaneously present).

These findings clearly indicate that both the 8–12 Hz peak and the 4–6 Hz peak are generated by an oscillating neurological loop of some kind, based probably on the stretch reflex and a pathway with a longer delay, respectively.

Muscle twitch times

It is just possible that muscle properties are altered in such a way that the parameters of the servo loop then cause it to oscillate at a lower frequency, although this seems most unlikely in view of the coexistence of oscillation in the two frequency bands. To check this, muscle twitch times were recorded after a sub-maximal contraction lasting for 1 h (and giving rise to 4–6 Hz tremor). No change was detected.

Conclusions

It is not yet possible to say how this slow tremor after prolonged sub-maximal contraction originates. It is apparently due to oscillation in a control system having a longer delay than the system involved in normal muscular tremor. This may mean that the secondary endings in muscle spindles with their less-marked dynamic properties are involved. They also have smaller, slower afferent fibres. Alternatively, tendon organs may be involved, or possibly long cortical loops might be implicated.

The facts that the tremor of Parkinson's disease behaves very much like this

---

FIG. 4. Sudden stretch of m. ext. dig. communis by solenoid and armature attached to middle finger. At arrow, stretch applied. Rectified and smoothed EMG averaged for 1.28 s afterwards and 32 sweeps, at 1913 ms intervals.

Record (a) shows a control experiment at the start of contraction (contraction sufficient to maintain finger raised against gravity). Record (b), taken 15 min after start, showing true summation of waveform at about 11 Hz, phase-locked to stretch. Record (c), taken at 45 min, shows large amplitude waves at 4 Hz summating (and superimposed waves at 11 Hz). See Lippold (1970) for a detailed description of the theory of this method.

This is strong evidence that the 4–6 Hz component of tremor after a prolonged submaximal contraction of a muscle is due to oscillation in a servo-loop involving muscle stretch receptors.

4–5 Hz tremor, and that patients typically have records like Fig. 4 line (c) and Fig. 3 line (c), direct attention towards the possibility that this fatigue tremor may prove to be a useful model for this disease.

Acknowledgements

These experiments were the joint work of the author, Jennifer Jessop and Sue Gottlieb.

I am indebted to The Wellcome Trust for a generous grant in support of this research.

REFERENCES

Bousfield WA 1932 Influence of fatigue on tremor. J Exp Psychol 15:104-107
Eshner AA 1897 A graphic study of tremor. J Exp Med 2:301-312
Furness P, Jessop J, Lippold OCJ 1977 Long-lasting increases in the tremor of human hand muscles following brief, strong effort. J Physiol (Lond) 265:821-831
Jessop J, Lippold OCJ 1977 Altered synchronization of motor unit firing as a mechanism for long-lasting increases in the tremor of human hand muscles following brief strong effort. J Physiol (Lond) 269:29P-30P
Lippold OCJ 1970 Oscillation in the stretch reflex arc and the origin of the rhythmical 8–12 c/s component of physiological tremor. J Physiol (Lond) 206:359-382
Lippold OCJ 1973 The origin of the alpha rhythm. Churchill-Livingstone, London
Stiles RN 1976 Frequency and displacement amplitude relations for normal hand tremor. J Appl Physiol 40:44-54

DISCUSSION

Merton: In 1967 we published experiments on tremor in which we concluded that, in our task, the 9 Hz tremor peak was the third harmonic of a fundamental of about 3 Hz, which we could sometimes see (Merton et al 1967). Do these results bear on yours?

Lippold: Yes, and I agree that it might be possible for the first harmonic of $4\frac{1}{2}$ Hz oscillation to appear as a 9 Hz peak in the frequency spectrum. What one then has to explain is where the $4\frac{1}{2}$ Hz fundamental has gone. In no practical system will you have large harmonics without the fundamental showing.

Wilkie: It is important to know that the effects of heavy muscular activity are so long-lasting. From the point of view of a neurological examination, where you ask patients to hold out their fingers as a check for tremor, you don't commonly ask them whether they were gardening the previous day, but I suppose you should.

Lippold: This enormous increase in tremor is seen in nearly all subjects after strong exertion or after having coffee to drink; smoking also enormously increases tremor at all frequencies.

Edwards: Are you dealing with a mechanism similar to or different from the Jendrassik manoeuvre used to reinforce a reflex?

Lippold: I think that Jendrassik's reinforcement is specific, and thus works on the 8–12 Hz peak alone, because its action is via the stretch reflex arc, but that this increase in tremor at all frequencies is a general phenomenon. Reinforcement increases tremor but only in the 8–12 Hz peak, called 'physiological tremor' (I am not sure why). In muscular fatigue, all frequencies of tremor are increased. That must mean that changes in the central excitatory state can't be the whole story. They may be responsible for part of the increase in tremor that we observe in fatigue, but clearly not all of it.

Merton: Is it not important to distinguish between this sort of increase in tremor and the more general kinds which are due to hormonal influences, such as adrenaline-induced tremor? You mentioned the effect of smoking. That is probably an adrenergically mediated effect. Is that tremor diminished by β-blocking agents?

Lippold: I don't know.

Merton: Is your accentuated tremor diminished by β-blockers?

Lippold: Normal 8–12 Hz muscular tremor is reduced by β-blockers but I don't know about the fatigue tremor.

Merton: So it might or might not be like adrenaline-induced tremor, which is abolished by these drugs (peripherally).

Lippold: The properties of muscle, and in particular the properties of muscle spindles, might be responsible for the second kind of tremor (4–6 Hz). This probably represents oscillation in a long loop, whereas the 8–12 Hz tremor is oscillation in a smaller loop with a shorter delay.

Edwards: Adrenaline or hyperthyroidism reduces the half-relaxation time by about 20%. Is a small change in the half-relaxation time from 100 ms to 80 ms sufficient to cause the alteration in the servo loop, to give the 4–6 Hz tremor?

Lippold: I think it's likely. The adrenaline-mediated tremor is seen if you have an arterial cuff and inject adrenaline into the arm distal to that. This is evidence that the increase in tremor is partly, if not entirely, due to changes in muscle properties, or at least that an entirely peripheral mechanism is involved.

Edwards: It might also be that the tremor is due to effects of adrenaline on sensory receptors, rather than to alterations in the mechanical properties of the muscle.

Lippold: It depends entirely what you mean by 'due to'. Adrenaline alters tremor by altering muscle properties. If, on the other hand, the mechanism giving rise to the 8–12 Hz tremor involves a whole servo loop, altering the properties of any part of the loop will alter the tremor. It is difficult to say that adrenaline tremor is actually 'due to' any single factor.

Stephens: There is still a controversy about the origin of the faster (8–12 Hz) peak in tremor. Your view is that it is oscillation in the stretch-reflex arc. Another view, for which there is also good evidence, is that this peak is due to motor neurons firing within that frequency band. In other words, it is simply a manifestation of the fact that motor units firing around 10 or 12 Hz have an unfused tetanic profile and therefore contribute a powerful component to tremor at these frequencies.

On the increase in the 4 Hz band, have you looked at spike trains of individual motor units for changes in the intervals between spikes that would have components in that 4 Hz band?

Lippold: We are doing this now. It will be very nice to see the underlying unit pattern. We have recorded the slow 4–6 Hz tremor using surface electrodes; there is good correlation between the tremor waveform and motor unit synchronization. We don't know how this is related to the firing of different kinds of units within the muscle.

On your first point, the evidence really doesn't support the view that the 8–12 Hz band has anything to do with properties of the motor neuron pool as such. You can cool the muscle, with the aim of altering the properties of the servo loop so that the delay is increased, which will lower the frequency. Cooling does this in two ways: when the properties of the muscle are altered the twitch time gets longer, which introduces an extra delay into the loop. Conduction time along the nerves increases and also introduces a delay. You can separate these effects. You can measure tremor from a hand muscle and cool the arm only; this changes only the nervous part of the loop. You can measure conduction velocity; that gives a measure of the increase in delay. You can then predict what the change in frequency ought to be, and experiment shows that it is changed by that amount. It is hard to explain that except as an oscillation in a loop. Tony Taylor's explanation of chance synchronization of motor units firing to give the frequency of normal tremor (Taylor 1962) doesn't fit these results. You can also introduce a mechanical or electrical input into the system, which will synchronize the tremor. In other words, you are synchronizing the oscillation in the loop. A mechanical step function introduced into a muscle during normal tremor synchronizes the tremor at 8 Hz, or whatever the actual tremor frequency in that subject might be. The oscillations produced behave exactly like the normal tremor. These oscillations are accompanied by action potentials, so this is nothing to do with the mechanical properties of the system. Lastly, you can open the loop, by de-afferentation, which abolishes the 8–12 Hz peak. We did this in the cat (Lippold et al 1959). You see the same effect in patients with tabes dorsalis, in some of whom the loop is effectively completely opened. In most of those patients the 8–12 Hz tremor is absent. Taken together, this all suggests that the 8–12 Hz tremor is due to oscillation in the stretch-reflex servo loop, in normal circumstances. There may be other causes, but they can't be major ones.

Stephens: During voluntary muscle contraction motor units fire at frequencies of 10–15 p.p.s. At these rates of firing each motor unit produces a force profile which is highly unfused (Taylor & Stephens 1976) and must therefore contribute power to the force spectrum in this frequency range. At least part of the peak in the power spectrum of tremor at 10 Hz must be attributable to this simple mechanical fact. The question is how much of the peak is produced by this mechanism.

Campbell: Two questions. First, in your second preparation (p 240), where the shift in frequency appeared quite suddenly, was there a change in motor performance at that time, as tested by maximum force or elicitable force? Secondly, you mentioned the illusion of movement in the first preparation (p 237). Are there any such illusions produced in the second preparation?

Lippold: On the first question, all we have measured is twitch times, and they seem to be unaltered. So in a general way, the muscle properties as elicited by a shock to the motor nerve are unaltered by sub-maximal contraction of the muscle lasting for about an hour.

On the second point, the sensation of movement in the fatigued finger is extraordinary. You can see your finger moving up and down with displacements of a centimetre or more and you have little control over this tremor. I imagine that something has gone wrong with the proprioceptive control. Even with full vision of the finger movements it is difficult to control the tremor unless you put your finger down, and then the tremor stops. But you cannot make an accurate movement at that stage.

Campbell: If you don't look at your finger, can you feel the tremor?

Lippold: You can feel the movement in terms of joint sensation but it is not as obvious as when you are looking at your finger.

Edwards: Is the explanation of this sensation analogous to Granit's explanation of why your arm rises after pressing the wrist hard against a wall? Most motor controls are of course unconscious, but he pointed out (Granit 1972) that when you alter relations between movement and force you can develop an error signal, of which you can become aware.

Lippold: This is a very similar sensation. You have a feeling that if you really tried hard you could do something about the slow tremor. In fact you probably can. With the arm-raising, equally, if you make a strong voluntary effort you can keep the arm down.

Edwards: Can Pat Merton tell us how long the reduction in cortical threshold lasts that follows movements of a single muscle in his experiments, described earlier?

Merton: As far as we are aware, when you stop contracting a muscle the reduction in its cortical threshold disappears forthwith.

Stephens: Some years ago Taylor and I tried to repeat Dr Merton's observations on the effects of vision on physiological tremor (Stephens & Taylor 1974). We could not find any effect of vision on the 10 Hz peak or any shift in the position of the peak when we introduced time delays between the recorded force signal and the signal displayed to the subject. I don't think these observations are established, therefore.

Merton: We (Dymott & Merton 1968) showed that depending on the position of the hand one could obtain tremor peaks at about 10 Hz which either were or were not abolished by closing the eyes. Unless the original experiments were closely repeated, therefore, there is no guarantee that visually sensitive peaks would be seen. The original experiments were carefully done by two separate groups and I regard them as firmly established (Sutton & Sykes 1967a,b, Merton et al 1967).

REFERENCES

Dymott ER, Merton PA 1968 Visually and non-visually determined peaks in the human tremor spectrum. J Physiol (Lond) 196:62P-64P

Granit R 1972 Constant errors in the execution and appreciation of movement. Brain 95:649-660

Lippold OCJ, Redfearn JWT, Vučo J 1959 The influence of afferent and descending pathways on the rhythmical and arrhythmical components of muscular activity in man and the anaesthetised cat. J Physiol (Lond) 146:1-9

Merton PA, Morton HB, Rashbass C 1967 Visual feedback in hand tremor. Nature (Lond) 216:583-584

Stephens JA, Taylor A 1974 The effect of visual feedback on physiological muscle tremor. Electroencephalogr Clin Neurophysiol 36:457-464

Sutton GG, Sykes K 1967a The effect of withdrawal of visual presentation of errors upon the frequency spectrum of tremor in a manual task. J Physiol (Lond) 190:281-293

Sutton GG, Sykes K 1967b The variation of hand tremor with force in healthy subjects. J Physiol (Lond) 191:699-711

Taylor A 1962 The significance of grouping of motor unit activity. J Physiol (Lond) 162:259-269

Taylor A, Stephens JA 1976 Study of human motor unit contractions by controlled intramuscular microstimulation. Brain Res 117:331-335

The pathophysiology of inspiratory muscle fatigue

PETER T. MACKLEM, CAROL COHEN, G. ZAGELBAUM and C. ROUSSOS

Meakins-Christie Laboratories, McGill University Clinic, Royal Victoria Hospital, Montreal, Canada H3A 1A1

Abstract The critical value of the rate of energy consumption of the inspiratory muscles above which fatigue occurs appears in some instances to be predictable from the relationship between energy demands and energy supplies rather than from the percentage of fatigue-resistant fibres in the inspiratory muscles. When this is the case the critical value of the external power produced by the inspiratory muscles is given by the product of muscular efficiency and the rate at which energy is supplied. Efficiency is reduced by hyperinflation and recruitment of the intercostal and accessory muscles of inspiration. The rate at which energy is supplied is decreased in states characterized by low cardiac output. The condition of low cardiac output, combined with the high oxygen cost of breathing against fatiguing loads, may be lethal in cardiogenic shock. Although the immediate cause of fatigue may not be related to reduced energy supplies, clinically useful predictions of conditions predisposing to fatigue result from an understanding of factors determining the balance between the energy demands and supplies of the inspiratory muscles. These predictions aid in the diagnosis of inspiratory muscle fatigue and have important therapeutic implications.

Although the exact mechanisms of skeletal muscle fatigue remain controversial, one can predict the conditions predisposing to inspiratory muscle fatigue from the hypothesis that fatigue will occur when the demand for energy of the working muscle exceeds the supply of energy. Some of these predictions can be tested experimentally. Thus in normal humans, hyperinflation, which shortens inspiratory muscle fibre length and decreases efficiency, and hypoxia, which decreases energy supplies, both lead to a reduction in endurance time (Roussos & Macklem 1977, Roussos et al 1979). In the dog, a reduction in cardiac output is sufficient to cause inspiratory muscle fatigue, respiratory arrest and death without any additional load to the inspiratory muscles. In

1981 Human muscle fatigue: physiological mechanisms. Pitman Medical, London (Ciba Foundation symposium 82) p 249-263

this experimental model death is prevented by artificial ventilation (Aubier et al 1980).

Because predictions based on the supply and demand of energy are supported by experiment, implications arising from the equation of Monod & Scherrer (1965) may be approximately correct even though the immediate cause of fatigue is not the depletion of intracellular high-energy phosphates. Thus:

$$C = \alpha + \beta.t_{\lim} \tag{1}$$

where C is the total energy consumed by a muscle from the beginning to the end of a fatiguing task, α is the energy consumed from stores with the muscle, β is the rate at which energy is supplied to the muscle by the blood, and $t_{\lim}$ is the endurance time.

Dividing equation 1 by $t_{\lim}$ yields

$$\dot{C} = (\alpha + \beta.t_{\lim})/t_{\lim} \tag{2}$$

where $\dot{C}$ is the rate of energy consumption by the muscle. But

$$\dot{C} = \dot{W}/E \tag{3}$$

where $\dot{W}$ is the external power produced by the muscle and E is its efficiency.

Substituting for $\dot{C}$ from equation 3 into equation 2 and solving for $t_{\lim}$ yields

$$t_{\lim} = \alpha/(\dot{W}/E - \beta) \tag{4}$$

By inspection of equation 4, as $(\dot{W}/E - \beta) \rightarrow 0$, $t_{\lim} \rightarrow \infty$.

The value for $(W/E - \beta)$ when $t_{\lim} = \infty$ is important physiologically and clinically because it defines the critical value of the external power $(\dot{W}_{crit})$ that a muscle can produce without becoming fatigued.

Alternatively, $\dot{W}_{crit}$ might be determined by the percentage of fatigue-resistant fibres in the muscle. If this hypothesis were correct, and the force–length relationships of fatiguable and fatigue-resistant fibres were the same, the critical force developed by a muscle, F_{crit}, expressed as a percentage of maximum force, should—within limits—not change as a function of either fibre length or energy supplies. For the respiratory system this means that the critical pressures developed by the respiratory muscles should be independent of lung volume and the rate at which energy is supplied.

Fig. 1 shows that this is indeed the case for the diaphragm in normoxic and hypoxic conditions. These graphs are plots of the inverse of endurance time at various transdiaphragmatic pressures (Pdi), expressed as a percentage of maximum Pdi (Pdi_{max}) in subjects breathing air and 13% O_2. The subjects breathed at functional residual capacity (FRC) through an inspiratory resistance and developed a Pdi with each inspiration that was a predetermined fraction of Pdi_{max}. They continued to do so as long as possible and when they

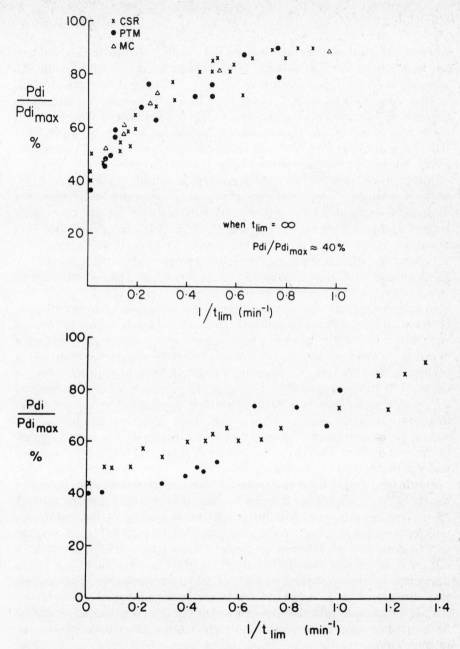

FIG. 1. *Upper:* relationship between strength of diaphragmatic contraction (Pdi/Pdi$_{max}$, trans-diaphragmatic pressure developed during each inspiration as a fraction of maximum trans-diaphragmatic pressure) and the inverse of diaphragmatic endurance time ($1/t_{lim}$) while subjects were breathing air. *Lower:* same relationship, subjects breathing 13% O$_2$. The intercept on the ordinate gives the critical value of Pdi when $t_{lim} = \infty$. (Roussos & Macklem 1977.)

reached the limit of their endurance, the experiment was terminated and endurance times were measured. The intercept on the ordinate is the Pdi when $t_{lim} = \infty$ and is thus the critical Pdi (Pdi_{crit}). The intercepts are identical at 40% Pdi_{max}. At all values greater than 40% Pdi_{max}, however, endurance times were decreased during hypoxia, except for very high values of Pdi of about 90% of maximum. These results are consistent with the hypothesis that the degree of hypoxia studied influenced only the fast twitch, slow-oxidative fibres, but not the fatigue-resistant ones.

Alternatively, in normoxic conditions there is a surplus in the supply of O_2 to the diaphragm. A certain degree of hypoxia is necessary before there is a decrement in performance and this depends not only on the delivery of oxygen to the working muscle but also on the muscle's metabolic rate. Thus hypoxia may not change the rate of O_2 uptake by a muscle performing a low rate of work, whereas the same degree of hypoxia might provide insufficient O_2 at a higher work rate. This could explain why hypoxia had no influence on Pdi_{crit} but reduced the endurance times at higher values of Pdi.

In experiments in which we examined the influence of hyperinflation, however, quite a different result was obtained, as shown in Fig. 2. Subjects were asked to breathe against a high inspiratory resistance and develop a negative pressure at the mouth (Pm), with each inspiration, that was a predetermined fraction of maximum negative mouth pressure (Pm_{max}) obtained at the appropriate lung volume. Experiments were done both at FRC and while subjects voluntarily maintained end-expiratory volume at half inspiratory capacity. Again, this is a plot of the inverse of endurance time against mouth pressure. However, in this experiment the critical mouth pressures fell from 50–60% of maximum at FRC to 24–35% of maximum under hyperinflated conditions.

Martin and Engel have shown in addition that hyperinflation produced merely by fixing respiratory frequency while subjects were breathing through an expiratory resistance, without any external loading of the inspiratory muscles, resulted in a finite endurance time (Martin et al 1978). Finally, the experiments in dogs referred to earlier (Aubier et al 1980), in which a reduction in cardiac output led to respiratory arrest without any force developed by the diaphragm in spite of persistent excitation, demonstrate that virtually all the fibres of the diaphragm become fatigued. These results are not compatible with the hypothesis that the critical workload or critical force development is determined merely by the percentage of fatigue-resistant fibres.

It therefore seems appropriate to examine the hypothesis that the critical power is determined by the point at which $\dot{W} = E\beta$. If this is the case, either a reduction in efficiency or a reduction in the rate at which energy is supplied will result in a decrease in $\dot{W}_{crit}$. Certainly hyperinflation or shortening

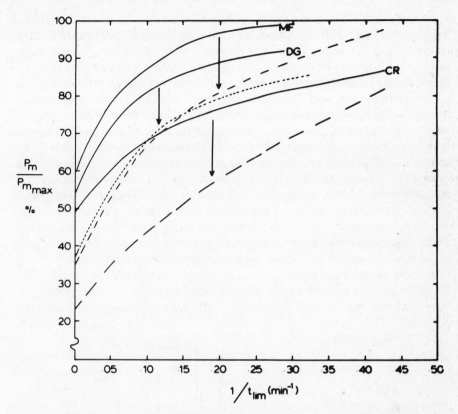

FIG. 2. Relationship between strength of inspiratory muscle contraction (Pm, negative mouth pressure developed during each inspiration while breathing against an inspiratory resistance) and the inverse of endurance time ($1/t_{lim}$). Solid lines are lines of best fit for three normal subjects breathing at FRC. Interrupted lines were obtained while subjects voluntarily maintained end-expiratory lung volume at half inspiratory capacity. Arrows indicate the influence of voluntary hyperinflation in each subject. (Roussos et al 1979.)

inspiratory muscle fibre length will markedly reduce efficiency and might explain, at least in part, the reduction in critical mouth pressure that was observed. Similarly, a reduction in cardiac output would be expected to reduce β. Thus the results of the hyperinflation experiments and the inspiratory muscle fatigue produced by a reduction in cardiac output are both explained by the hypothesis that the critical power output of the inspiratory muscles is determined by the product of efficiency and the rate of energy supply.

If this is true it becomes pertinent to examine the factors that determine efficiency and the distribution of energy supplies to the inspiratory muscles as against the rest of the body tissues. One of the factors influencing efficiency has already been mentioned, namely initial fibre length as determined by lung volume. Another is the pattern of breathing. When the diaphragm alone contracts it displaces both the rib cage and the abdomen as its fibres shorten. When the intercostal muscles contract alone, the rib cage is inflated but the reduction in pleural pressure is transmitted across the flaccid diaphragm to the abdomen, resulting in an inward expiratory displacement of that compartment. This lengthens the fibres of the diaphragm. When the intercostal muscles and the diaphragm contract simultaneously, so that there is no abdominal displacement, the diaphragm contracts quasi-isometrically. In these circumstances the diaphragm does no external work yet continues to consume considerable energy. Its efficiency is zero. Thus any load which leads to recruitment of the intercostal and accessory muscles of inspiration results in a decrease in efficiency which will theoretically reduce the critical force or workload that leads to inspiratory muscle fatigue.

We have measured total-body O_2 consumption on normal subjects at rest and while breathing against a fatiguing inspiratory load. The results are shown in Fig. 3. Total-body O_2 consumption increased by a factor of 2.75. This confirms the findings of Bradley & Leith (1978) indicating that the

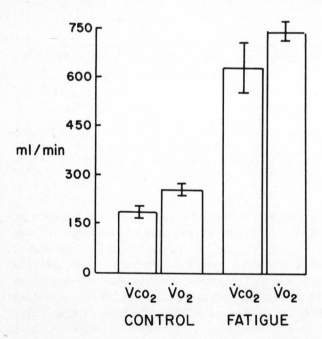

FIG. 3. Total O_2 consumption ($\dot{V}_{O_2}$) and CO_2 production ($\dot{V}_{CO_2}$) in normal man at rest (control) and while doing fatiguing inspiratory work (fatigue). (Macklem 1980.)

oxygen cost of breathing may become very great, even exceeding one litre of O_2 per minute, when the inspiratory muscles do fatiguing work. Furthermore, the O_2 cost of breathing in chronic airflow limitation increases disproportionately as ventilation increases by comparison with normal subjects (Campbell et al 1957, Cherniack 1959), as illustrated in Fig. 4. Also shown is

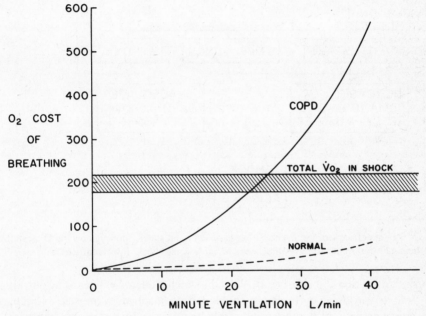

FIG. 4. Diagrammatic representation of the O_2 cost of breathing in normal subjects and in patients with chronic obstructive pulmonary disease (COPD) (Campbell et al 1957, Cherniack 1959) as a function of minute ventilation. The shaded area shows values for total O_2 consumption ($\dot{V}o_2$) in septic shock (MacLean et al 1967). (Macklem 1980.)

the total-body O_2 consumption measured in septic shock by MacLean et al (1967). When the O_2 cost of breathing in these circumstances is compared to total-body O_2 consumption in conditions which lead to impaired O_2 delivery, the interesting fact emerges that the O_2 demands of the inspiratory muscles may equal the O_2 supply to the whole body.

This may be particularly important clinically, both in septic shock and in conditions characterized by low cardiac output. The sequence of events for conditions with a low fixed cardiac output is shown schematically in Fig. 5. The combination of inadequate delivery of oxygen to the tissues with an excessive demand for it by the inspiratory muscles is presumably lethal and suggests that inspiratory muscle fatigue may be the cause of death. Indeed,

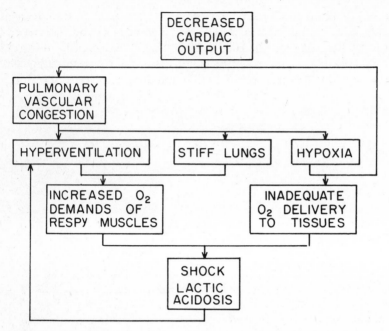

FIG. 5. Scheme of sequence of events in low cardiac output states, showing how the O_2 cost of breathing may contribute to the pathogenesis of cardiogenic shock. (Macklem 1980.)

experiments in dogs (Aubier et al 1980), in which a decrease in cardiac output sufficient to produce cardiogenic shock led to inspiratory muscle fatigue, respiratory arrest and death, support this hypothesis. Furthermore, artificial ventilation in a similar group of dogs, with an equal reduction in cardiac output, prevented death (Aubier et al 1980).

The therapeutic implications of this line of reasoning are clear. In the face of inadequate oxygen delivery to the tissues, from whatever cause, there is a serious risk of inspiratory muscle fatigue. If the clinical manifestations of fatigue become apparent, steps should immediately be taken to unload the inspiratory muscles. If this cannot be done by decreasing airway resistance or increasing compliance, serious consideration should be given to artificial ventilation.

To finish, we should like to present an anecdotal case history supporting these ideas. The patient was a 54-year-old man with myocardial infarction, hypotension, mild pulmonary oedema but essentially normal arterial blood gas tensions, breathing supplementary oxygen. A few hours later, although his condition was thought to be stable, the house officer noted that intermittently there were inward abdominal displacements during inspiration. In the course of two or three hours these paradoxical abdominal displacements became persistent. This clinical sign indicates weak diaphragmatic contraction so that the fall in pleural pressure during inspiration is transmitted across the diaphragm to the abdomen. The sign has been reported only in diaphragmatic paralysis (Newsom Davis et al 1976) and diaphragmatic fatigue (Macklem 1980). A further arterial blood gas measurement revealed that the patient had developed previously unsuspected hypercapnia and acute respiratory acidosis. As he was already receiving optimal treatment for heart failure, artificial ventilation was instituted. This stabilized his condition, his heart failure improved, and after a few days the patient was extubated and subsequently discharged from hospital. Admittedly, one case does not make a textbook, but nevertheless we propose that this patient's respiratory muscles were probably the most metabolically active tissues in his body. In the face of an inadequate oxygen supply system their energy demands put an extra load on the heart, which was unable to supply sufficient oxygen. Accordingly the patient developed inspiratory muscle fatigue and acute hypercapnic respiratory failure. We suggest that artificial ventilation, by putting his respiratory muscles to rest, did three things.

(1) It allowed his inspiratory muscles to recover from fatigue.

(2) It maintained normal alveolar ventilation.

(3) It decreased the energy demands of the respiratory muscles, diverting needed energy to other tissues.

It would therefore seem likely that the patient's subsequent recovery and discharge from hospital were substantially aided by putting his respiratory muscles to rest.

In conclusion, it appears reasonable that an understanding of the factors determining the balance between the energy supplies and demands of the inspiratory muscles leads to clinically important predictions of conditions predisposing to inspiratory muscle fatigue and acute hypercapnic respiratory failure. These predictions aid in the diagnosis of inspiratory muscle fatigue and have important therapeutic implications.

Acknowledgements

The work reported in this paper was supported by grants from the Medical Research Council of Canada. C. Roussos is a scholar of the Medical Research Council of Canada.

REFERENCES

Aubier M, Trippenbach T, Sillye G, Viires N, Roussos CS 1980 Respiratory failure in cardiogenic shock. Proc Int Union Physiol Sci 14:305 (abstr)

Bradley ME, Leith DE 1978 Ventilatory muscle training and the oxygen cost of sustained hyperpnea. J Appl Physiol 45:885-892

Campbell EJM, Westlake EK, Cherniack RM 1957 Simple methods of estimating oxygen consumption and efficiency of the muscles of breathing. J Appl Physiol 11:303-308

Cherniack RM 1959 The oxygen consumption and efficiency of the respiratory muscles in health and emphysema. J Clin Invest 38:494

Macklem PT 1980 Respiratory muscles: the vital pump. Chest 78:753-758

MacLean LD, Mulligan WA, McLean APH, Duff JH 1967 Patterns of septic shock in man: a detailed study of 56 patients. Am J Surg 166:543-558

Martin JC, Habib M, Roussos CS, Engel LA 1978 Inspiratory muscle activity during induced hyperinflation. Physiologist 21:77 (abstr)

Monod H, Scherrer J 1975 The work capacity of a synergic muscular group. Ergonomics 8:329

Newsom Davis J, Goldman M, Casson M 1976 Diaphragm function and alevolar hypoventilation. Q J Med 45:87-100

Roussos CS, Macklem PT 1977 Diaphragmatic fatigue in man. J Appl Physiol 43:189-197

Roussos CS, Fixley M, Cross D, Macklem PT 1979 Fatigue in inspiratory muscles and their synergic behavior. J Appl Physiol 46:897-904

DISCUSSION

Wilkie: It is obviously important to have shown that the work of breathing can become so great that it exceeds the amount of oxygen taken in. Dr Macklem also wrote some equations about the total supply of 'energy' to a muscle. Let me make a general plea here that a clear distinction should always be made between physical quantities, like heat and work, and chemical quantities. The blood doesn't supply 'energy' as such to the muscles; it supplies chemicals. Some very bad mistakes have resulted from failure to follow this rule; some of them are enshrined in the recommendations of the Commission set up by the International Union of Physiological Sciences on units and symbols for heat-balance studies (Bligh & Johnson 1973; see comments by Wilkie 1975). Thus it is no wonder that papers continue to be published where good experimental work is marred by false interpretation, largely because the authors talk about energy supply when they mean supply of chemicals. This is not mere pendantry. As a simple example showing where the two can be different, consider the utilization of glycogen. This can end up either as CO_2 and water if it is oxidized, or as lactate if it isn't. The amount of physical energy available for doing work is quite different in the two cases. One must distinguish between physical and chemical quantities.

Edwards: I would support that. The analysis of Monod & Scherrer (1965) may be intuitively correct but it does not add much to the understanding of fatigue. There is a danger in using mathematical symbols when it is not possible to define accurately the components of the equation.

Macklem: I used the equation to develop the idea that the efficiency, inspiratory muscle blood flow and arterial oxygen content were critically important in determining the conditions in which inspiratory muscle fatigue is likely.

Moxham: I was interested in the study in which hypoxia shortened endurance times but did not influence the critical level of transdiaphragmatic pressure (Pdi). I still find that difficult to understand.

Macklem: A possible explanation is that hypoxia didn't affect the critical level because the fatigue-resistant (slow twitch) fibres were not influenced by that degree of hypoxia. An alternative explanation is that for that degree of force developed by the diaphragm, there was no reduction in the amount of oxygen supplied to the diaphragm. After all, in normal circumstances the diaphragm does not take up all the oxygen in the blood that perfuses it, so a surplus of oxygen is being supplied. If so, in order to influence the performance of the diaphragm by hypoxia one would have to produce a particular level of hypoxia before the oxygen supply became inadequate.

Campbell: How do you explain the earlier parts of the curve?

Macklem: The diaphragm is producing more force at higher levels of Pdi, in which circumstances the particular level of hypoxia we used may have resulted in an inadequate oxygen supply.

Campbell: In the experiment on lung volume, did you have diaphragmatic pressure observations? It is not just the switch from one muscle to another set of muscles?

Macklem: No. In these experiments, where you develop a mouth pressure and can choose the muscles that are going to develop that pressure, there is an alternation, in which you contract the diaphragm forcefully for a few breaths and then recruit the intercostal muscles and contract those. You alternate between the two sets of muscles. In the hyperinflation experiments there is still that alternation. There is also sustained contraction of mostly the intercostal muscles to maintain that lung volume.

Campbell: May I make three points? First, I don't think anybody would quibble with the conceptual scheme that if the oxygen cost of breathing goes beyond a certain point, the respiratory muscles consume oxygen that should be used elsewhere. I am however concerned about the 'numbers' available. You quoted our observations on the work of breathing (Campbell et al 1957). You can't measure the oxygen consumption of respiratory muscles in the way you can for most organs because you can't apply the Fick principle. You have to do it by measuring oxygen consumption at various levels of breathing and subtracting the measurements from each other, assuming that nothing else has changed. Having done such experiments I came to the conclusion that when ventilation was increased there was increasing oxygen consumption in other sites than the respiratory muscles.

Secondly, normal subjects who are asked to hyperventilate maximally can sustain around 200 l/min for about 30 seconds and then the ventilation falls to around 120l/min. They can keep that up for a long time (10–30 min), so it looks like a fast frequency type of fatigue. In patients with chronic airflow obstruction, who have the biggest work of breathing problem, the maximum breathing capacity can be sustained for at least four minutes, suggesting that it isn't fatigue of the respiratory muscles which limits the amount they can breathe in these circumstances (Clark et al 1969).

Thirdly, I agree with you and Professor Roussos that in septic shock the combination of increased respiratory work with decreased cardiorespiratory ability to supply O_2 and remove CO_2 must lead to a vicious spiral. I have long thought that this combination may be an agonal event but had not expected it to develop sufficiently slowly or detectably for artificial ventilation to be practically useful. I hope you will define the indications more precisely before artificial ventilation is applied over-enthusiastically.

Macklem: I agree that the oxygen cost of breathing may not represent the oxygen cost of contracting the inspiratory muscles. Patients with difficulty in breathing, and subjects breathing against inspiratory loads, are frequently grasping the arms of their chairs and tensing up other muscles, which are all consuming oxygen. However, this *is* the oxygen cost of breathing. If you were to take over their ventilatory effort they would relax these other muscles too.

I am obviously reluctant to suggest artificial ventilation for trivial conditions. I have always thought that artificial ventilation is used too frequently and in trivial circumstances. But I have wondered about cardiogenic shock. The fatality rate is high. A trial of artificial ventilation here is probably warranted, to see whether it does improve the survival rate.

I am surprised that the maximal sustainable ventilation was so high in patients with chronic airflow obstruction. In patients with chronic bronchitis and emphysema whom they asked to hyperventilate, Dr Richard Pardy and Dr Roussos (personal communication) found electromyographic evidence of fatigue. We also found EMG evidence of fatigue of the diaphragm in such patients when they were being exercised.

Campbell: Perhaps these patients don't use their diaphragms very much during these manoeuvres.

Edwards: We brought respiratory muscle fatigue into this symposium not just because it is of interest to respiratory physiologists and physicians, but because it can be thought of as a special example of skeletal muscle function. To begin with, we have to look carefully at what we mean by efficiency— mechanical efficiency, or energetic efficiency—and at the mechanisms underlying fatigue. Studies of respiratory muscle function are of considerable interest to those of us who are normally more concerned with fatigue developing in one dimension—that is, force along the line of the muscle length. Study of respiratory muscle fatigue gives us the opportunity to see the effects of chronic activity, by analogy with the effects on the heart of chronic failure. The same changes are being seen in respiratory muscles. The consequences, for example, of lungs that are chronically enlarged include chronically reduced fibre lengths in the diaphragm. If a muscle is allowed to shorten for a long period it sheds sarcomeres to optimize sarcomere length (Goldspink et al 1974).

A question arises in a patient with chronic airways obstruction of the extent to which you see damage of the respiratory muscles. Dr Richard Hughes and Dr Vinod Sahgal in Chicago (unpublished) have obtained diaphragm muscle biopsies at operation from patients with chronic obstructive lung disease. These show some of the features of a myopathy. It is relevant to ask whether an 'over-use' myopathy can develop in respiratory muscles that may contribute to, or predispose to, the development of fatigue in these muscles.

Morgan-Hughes: We looked at intercostal muscle biopsies from patients with chronic obstructive airway disease who were undergoing thoracic surgery and found a high incidence of abnormalities. The changes were largely non-specific but in some cases there was a high proportion of ragged red fibres (Olson et al 1972). The cause of these changes is uncertain, but I suppose they could arise from chronic hypoxia.

Campbell: Gertz et al (1977) found that the concentrations of ATP and creatine phosphate were low in both intercostal and quadriceps muscles in patients with respiratory failure due to chronic lung disease. The concentrations were particularly low in the quadriceps, suggesting that the changes in the respiratory muscles are due to a general disturbance rather than simply to increased respiratory work.

Edwards: One interpretation of those findings is that the muscles are chronically hypoxic. Another interpretation is that these are muscles that are working hard. In the light of this, you have to compare the metabolite content of these muscles not with resting muscle, but with normal muscle working at a comparable rate, which might have a similar metabolic profile to fatigued muscle.

Saltin: I agree. The phosphagen concentration may be an expression of how intensely involved the muscles are in hard exercise.

Campbell: In such patients with respiratory muscle failure, whether it is due to hypoxia or to malnutrition (some of them are malnourished and dehydrated), there could be a general effect on muscle function which of course affects respiratory muscles as well. That would be another feedback loop.

Edwards: Eric Hultman's group have looked at high energy phosphates and lactate in acutely and chronically ill patients.

Hultman: We studied a series of severely ill patients with circulatory and respiratory insufficiency, most of them being treated with artificial respiration for shorter or longer periods (Bergström et al 1976). The acutely ill patients showed an increase in muscle lactate content and a decrease in the phospho-creatine and ATP stores in the skeletal muscle. The lactate increase and phosphocreatine decrease could be explained by a relative hypoxia in the muscle but the decrease in ATP and in total adenine nucleotide content must have an additional explanation. The decrease in the adenylate pool was even more pronounced in patients with chronic diseases. No accumulation of lactate in muscle was observed in these patients but the mean ATP level was depressed to 50% of normal. The low adenine nucleotide level in muscle tissue is thought to be due to the increased formation and deamination of AMP during hypoxia, in combination with a decreased rate of purine synthesis in the liver and/or a decreased capacity for 'purine salvage' in the muscle. This might be explained by a low energy state in muscle or liver, or be due to other metabolic disturbances or tissue damage in these severely ill patients. Similar changes might be responsible for the observed findings in respiratory muscle.

Roussos: Were the non-specific abnormalities in respiratory muscle rever-sible, with any kind of training for instance?

Morgan-Hughes: We have no information about that.

Edwards: If such changes are due to over-use, one wouldn't expect them to improve with training.

Roussos: In that case, perhaps they need rest, as was said earlier.

Moxham: It struck me, in the paper by J. A. Campbell et al (1980) showing non-specific morphological changes in the respiratory muscles of patients undergoing thoractomy for suspected tumours, that these patients really had quite good lungs from the point of view of airways obstruction, with an FEV_1 of 64% of the predicted value. So the respiratory muscles of these patients may not have been particularly overactive.

REFERENCES

Bergström J, Boström H, Fürst P, Hultman E, Vinnars E 1976 Preliminary studies of energy-rich phosphagens in muscle from severely ill patients. Crit Care Med 4:197-204

Bligh J, Johnson KG 1973 Glossary of terms for thermal physiology. J Appl Physiol 35:941-961

Campbell EJM, Westlake EK, Cherniack RM 1957 Simple methods of estimating oxygen consumption and efficiency of the muscles of breathing. J Appl Physiol 11:303-308

Campbell JA, Hughes RL, Sahgal V, Frederiksen J, Shields TW 1980 Alterations in intercostal muscle morphology and biochemistry in patients with obstructive lung disease. Am Rev Respir Dis 122:679-686

Clark TJH, Freedman S, Campbell EJM, Winn RR 1969 The ventilatory capacity of patients with chronic airways obstruction. Clin Sci (Oxf) 36:307-316

Gertz I, Hedenstierna G, Hellers G, Wahren J 1977 Muscle metabolism in patients with chronic obstructive lung disease and acute respiratory failure. Clin Sci Mol Med 52:395-403

Goldspink G, Tabary C, Tabary JC, Tardieu C, Tardieu G 1974 Effect of denervation on the adaptation of sarcomere number and muscle extensibility to the functional length of the muscle. J Physiol (Lond) 236:733-742

Monod H, Scherrer J 1965 The work capacity of a synergic muscular group. Ergonomics 8:329

Olson W, Engel WK, Walsh GO, Einaugler R 1972 Oculocraniosomatic neuromuscular disease with 'ragged-red' fibers. Arch Neurol 26:193-211

Wilkie DR 1975 Thermodynamic errors by International Commission. Nature (Lond) 257:87-88

Fatigue in human metabolic myopathy

C. M. WILES*, D. A. JONES and R. H. T. EDWARDS

Department of Human Metabolism, The Rayne Institute, University College London School of Medicine, London WC1 6JJ, UK

Abstract The ability of muscle fibres to sustain force can be related to their economy of energy utilization and to their capacity to regenerate energy under the prevailing conditions (aerobic or anaerobic) of contraction. The pathophysiology of muscle fatigue is analysed in patients with thyroid dysfunction and with impaired glycogenolysis, and in a patient with abnormal mitochondrial function.

Muscle from hypothyroid patients, like cooled muscle, is slow in relaxing and shows a reduced energy requirement (energy economy) and reduced fatiguability, whereas muscle of hyperthyroid patients may show the opposite features. In myophosphorylase deficiency the energy economy is normal in the fresh state and increases as contraction proceeds; however, fatigue is premature and associated with impaired excitation rather than an overall depletion of energy stores.

With abnormal mitochondrial function the muscle tends to be effectively anaerobic and fatigue is associated with impaired excitation–contraction coupling. This appears to result from either muscle ischaemia or the dominant use of anaerobic metabolism for energy regeneration.

Fatigue in these disorders of energy metabolism may ultimately be due to a reduced supply of ATP but direct evidence of this is lacking and, if it occurs, its physiological expression is probably variable.

Patients with many types of muscle disorder commonly describe symptoms that are collectively known as 'fatigue'. Progressive weakness or heaviness of the limbs on exertion, muscular stiffness or aching and generalized tiredness or breathlessness are typical of such symptoms and should be distinguished from 'fixed' weakness present at the onset of activity. This latter feature is more likely to be described in terms of difficulty in rising from a low chair, in raising the arms to comb the hair, or in gripping objects with the fingers. Thus weakness is a failure to develop the required or expected force, whereas

---

Present address: The National Hospital for Nervous Diseases, Queen Square, London WC1N 3BG, UK

1981 Human muscle fatigue: physiological mechanisms. Pitman Medical, London (Ciba Foundation symposium 82) p 264-282

fatigue is an inability to sustain it. Ultimately, processes giving rise to fatigue may, if profound, give rise to fixed weakness—a situation common in myasthenia gravis, for example. But in discussing the pathophysiology of symptoms it remains valuable to distinguish between these two aspects of a patient's symptoms.

In any muscle disorder a variety of factors (Table 1) may potentially

TABLE 1 Factors associated with the symptom of fatigue in myopathy

1. Increased recruitment and firing rate of motor units due to:
   (a) reduced complement of muscle fibres, or
   (b) muscle fibre fatigue
2. Altered skeletal stability and mechanical advantage
3. Muscle pain
4. Cardio-respiratory responses to increased effort
5. Cardiac and respiratory muscle weakness
6. Secondary metabolic events resulting from abnormal muscle metabolism
7. Altered afferent input from muscle

contribute to the patient's feeling of 'fatigue'. This paper is particularly concerned with examining muscle fibre fatigue—the ability of muscle fibres to sustain force in response to adequate excitation by the motor nerve. Impairment of this particular function of muscle fibres can be expected to be a particular feature of 'metabolic' muscle diseases as opposed to the fixed weakness which dominates the clinical picture in dystrophy and inflammatory myopathy as a result of frank replacement and loss of contractile elements. In Table 2 potential sites in the contractile sequence are shown where abnormalities might have a limiting effect on force generation. Fatigue may be due to

TABLE 2 Sites of muscle fibre fatigue in myopathy

1. Neuromuscular transmission (myasthenia gravis)
2. Sarcolemmal excitation (myotonia)
3. Excitation–contraction coupling (dantrolene treatment, acidosis)
4. Energy economy:
   (a) hypermetabolic myopathy
   (b) thyroid disease
5. Energy generation:
   (a) myophosphorylase and phosphofructokinase deficiencies
   (b) mitochondrial abnormalities:
       substrate utilization or transport
       respiratory chain defects
6. Contractile processes

impaired excitation, when either the neuromuscular junction (in myasthenia gravis) or the sarcolemmal membrane and T (transverse) tubular system (in myotonia congenita) cannot sustain physiological rates of impulse transmission. Evidence of failure of excitation–contraction coupling has been found after ischaemic fatigue in normal muscle (Edwards et al 1977a), in mitochondrial disorders (De Jesus 1974, Morgan-Hughes et al 1977) and after treatment with drugs such as dantrolene (a muscle relaxant sometimes used in the treatment of spasticity). Since muscle is in essence a machine for transducing chemical energy into force, work and heat, disturbances of energy metabolism might be expected to predominate as causes of muscle fatigue. In 1951 McArdle described the clinical and metabolic features of myophosphorylase (EC 2.4.1.1) deficiency and since then an increasingly large number of biochemical lesions in energy-yielding pathways have been described. In this paper we describe some studies of muscle contractility and energy turnover in such disorders in an attempt to correlate the pathophysiology of muscle fatigue with specific biochemical defects. In particular we analyse the cause of fatigue in patients with altered thyroid function, in a group with impaired glycogenolysis and in a single patient with abnormal mitochondrial function of the skeletal muscle.

Methods and patients

All our studies have been of isometric contractions of adductor pollicis or quadriceps femoris, using the apparatus described by Edwards et al (1977b). Contractions have been either voluntary or electrically stimulated and, in the latter case in adductor pollicis, muscle excitation has been assessed from the amplitude of the evoked muscle action potential after supramaximal stimulation of the ulnar nerve at the wrist. Relaxation rate has been measured from the differentiated force record (Wiles et al 1979) and the frequency–force response of the muscle ascertained using a programme of electrical impulses at 1, 10, 20, 50 and 100 Hz (Edwards 1978). Energy turnover has been measured in two ways. In the first, needle muscle biopsies (Edwards et al 1980) are taken before and after fatiguing ischaemic isometric contractions of the quadriceps muscle and the energy turnover is measured in terms of the ATP hydrolysed during the contraction as calculated from the change in muscle content of ATP, phosphoryl creatine and lactate (Edwards et al 1975). The second technique utilizes a thermocouple or thermistor probe inserted into the muscle to measure the rate of metabolic heat production, which is in turn an estimate of the rate of energy turnover when the muscle is ischaemic and doing little or no external work (Edwards et al 1975, Wiles 1980).

The hypo- and hyperthyroid patients all had unequivocal biochemical evidence of altered thyroid status but had not necessarily presented with muscular symptoms. Four patients had myophosphorylase deficiency (proved by biochemical and histochemical measurement of the enzyme) and one patient had similarly proved phosphofructokinase deficiency. The third category consists of a single male patient studied in collaboration with Dr J. A. Morgan-Hughes. This patient had profound exercise intolerance and fatiguability with a marked tendency to develop lactic acidosis and a low maximal oxygen uptake of about 1 l/min. Biochemical studies showed a reduced capacity of mitochondria to oxidize NAD-linked substrates.

Procedures on patients were performed after obtaining their informed consent. All procedures were approved by the Research Ethics Committee of the Royal Postgraduate Medical School and Hammersmith Hospital or the Committee of Ethics of Clinical Investigation at University College Hospital.

Results and discussion

The energy economy

When fatigue is ascribed to a deficit of cellular ATP an important consideration is whether the energy requirement for a given force—that is, the energy economy—is normal. If normal muscle is cooled the rate of intramuscular temperature rise (dT/dt) in a maximum contraction and the ATP turnover rate for a given submaximal force are both reduced (Edwards et al 1972, Edwards & Wiles 1980a, b). In muscle of hypothyroid patients dT/dt in maximum contractions is $0.53 \pm 0.08\,°C\ min^{-1}$ (mean $\pm$ SEM, $n=4$) compared to the normal range of $0.91 \pm 0.28\,°C\ min^{-1}$ (mean $\pm$ 2SD, $n=48$). The average rate of turnover of ATP in submaximal contractions sustained to fatigue is $6.7 \pm 1.5\,\mu$mol ATP s^{-1} mmol^{-1} total creatine (mean $\pm$ SEM, $n=5$) in hypothyroid patients, compared to normal values of $26.0 \pm 3.4\,\mu$mol ATP s^{-1} mmol^{-1} total creatine (mean $\pm$ SEM, $n=6$, $P<0.001$) (Wiles et al 1979).

Both cooling and hypothyroidism are associated with slowing of relaxation. The contractile speed of muscle fibres has long been associated with the energetic economy. Fig. 1 shows that there is a general relationship between the rate of energy turnover per unit force and contractile speed as indicated by relaxation rate. Thus the suggestion by Lambert and coworkers in 1951 that the slowness of hypothyroid muscle was connected 'to a decrease in the rate of energy liberation' has proved well founded.

An improvement in the economy of force maintenance (i.e., less energy is required for a given force) should result in reduced fatiguability, while a reduction in its economy should result in a greater susceptibility to fatigue. This prediction is borne out in practice. Cooled normal muscle (Edwards et al 1972, Edwards & Wiles 1980b) and hypothyroid muscle (Wiles et al 1979) can sustain force for longer before fatigue than warmed or euthyroid muscle (see also Fig. 2). Hyperthyroid muscle, on the other hand, is more fatiguable, which fits in with its high rate of turnover of ATP (Wiles et al 1979). We

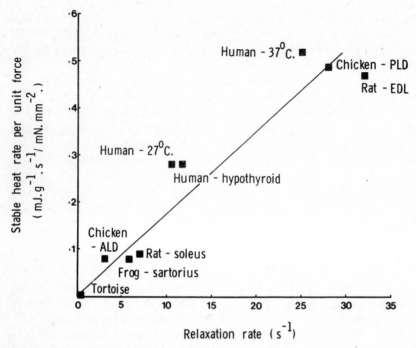

FIG. 1. The relationship between stable metabolic heat production and relaxation rate in various species, including man, is shown. The rate of stable metabolic heat production ($mJ g^{-1} s^{-1}$) in a fused tetanus has been divided by the isometric force production per unit cross-section ($mN mm^{-2}$). Relaxation rate is expressed as the rate constant (s^{-1}) for the exponential phase of relaxation from an isometric contraction. The temperatures at which measurements were made were: tortoise, 0 °C; frog sartorius, 0 °C; chicken anterior latissimus dorsi, posterior latissimus dorsi (ALD, PLD), 20°C; rat soleus, extensor digitorum longus (EDL), 27°C; human (hypothyroid), 36 °C and human (normal thyroid status), 27 and 37 °C. (For original references and results see Wiles 1980, or available on request.)

propose, therefore, that the resistance to fatigue shown by cooled or hypothyroid muscle is primarily a function of its improved energy economy.

Clear alterations in the energy economy are probably uncommon. The two patients with 'hypermetabolic' myopathy described by Luft et al (1962) and Haydar et al (1971) both appear to have had excessive fatiguability. In both patients rates of mitochondrial respiration were maximal in the absence of a phosphate acceptor and the respiratory control ratio was near unity. In muscle from the second case, Di Mauro et al (1976) found evidence that continual recycling of calcium ions between mitochondria and sarcoplasm could result in sustained high rates of respiration. Energy turnover was therefore excessive in relation to the needs of the muscle and could have contributed to fatiguability. Patients have also been described with myopathy

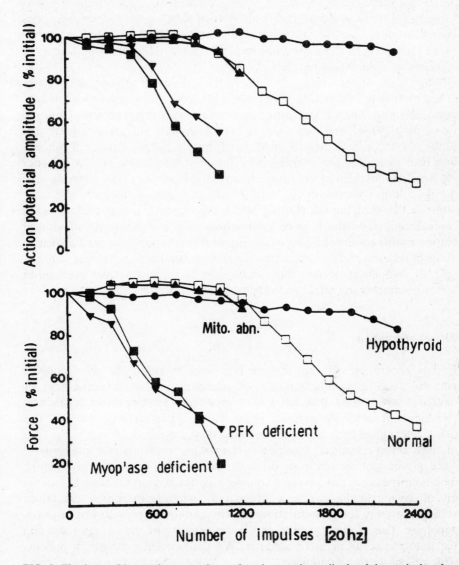

FIG. 2. The force of isometric contraction and peak-to-peak amplitude of the evoked surface motor action potential in adductor pollicis are related to the number of supramaximal stimuli given to the ulnar nerve at the wrist at 20 Hz. The circulation to the arm is occluded both during the contraction and for 3 min before it. □, normal subject (male, 29 yr); ●, hypothyroid patient (female, 73 yr); ■, myophosphorylase-deficient patient (male, 16 yr); ▼, phosphofructokinase-deficient patient (male, 27 yr); ▲, patient with mitochondrial abnormality (male, 20 yr).

presenting primarily as weakness, in association with mitochondrial respiratory control ratios of close to unity but normal basal metabolic rates (e.g. van Wijngaarden et al 1967). In such patients the regulation of metabolic rate to the requirements of the contracting muscle is likely to be poor, but the energy economy of contraction normal.

When the regeneration of ATP is limited, as in myophosphorylase-deficient muscle contracting under ischaemic conditions, alterations in economy may play a part in protecting against fatigue. In such patients (Wiles & Edwards 1979) we find that dT/dt in brief maximum contractions is normal ($0.98 \pm 0.08\,°C\,min^{-1}$, mean $\pm$ SEM, $n = 4$) but that in a submaximal contraction held to fatigue the average ATP turnover rate is low ($10.5 \pm 2.9\,\mu mol$ ATP s^{-1} $mmol^{-1}$ total creatine, mean $\pm$ SEM, $n = 4$). This suggests that during a long contraction the rate of energy turnover declines. We were interested in studying the slowing of relaxation which occurs during such a contraction, particularly since contracture is a commonly noted clinical feature in this condition. Excess slowing of relaxation does indeed appear to occur in relation to the force × time sustained, compared to normal subjects (Fig. 3). We speculate that this can be seen as a compensatory mechanism which maximizes the effect of available energy resources.

Energy generation

Energy generation may fail if there is a block in either the glycogenolytic pathway (e.g. myophosphorylase or phosphofructokinase deficiency) or oxidative metabolism. Whether a block produces a limiting effect depends on the dominant metabolic pathway in use during a particular contraction, this being determined by factors such as blood flow, oxygen delivery and the type of fibre being recruited. The two pathways are evidently not alternatives, since glycogenolysis tends to predominate in conditions where oxidative metabolism cannot proceed (e.g. during high force contractions when delivery of oxygen to the muscle is attenuated), whereas in aerobic conditions sufficient energy for long-duration exercise cannot be provided from glycogenolysis. This is due partly to the lower capacity of the energy store and partly to the resulting lactic acidosis. We shall consider fatigue in patients with a block in these pathways.

Myophosphorylase and phosphofructokinase deficiency. In ischaemic conditions muscle deficient in these enzymes fatigues more rapidly than normal (Fig. 2, p 269; Edwards & Wiles 1980a). Force fatigue in these muscles is always associated with a rapid decline in the amplitude of the action potential, the recovery of which appears to depend on restoration of the circulation.

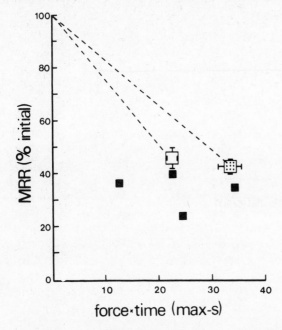

FIG. 3. The reduction in maximum relaxation rate (MRR) after ischaemic contractile activity of adductor pollicis in normal and myophosphorylase-deficient subjects. MRR is determined from the differentiated force record in an isometric tetanus (Wiles et al 1979) and expressed as % of the value in fresh muscle. After control measurements and a 3-min period of preliminary ischaemia, to reduce stores of oxygen, a standard amount of contractile activity is performed (measured as the force–time integral ÷ maximum tetanic force at 100 Hz, i.e. equivalent number of seconds at maximum force, or max-s) and MRR is again measured. Normal subjects performed two patterns of ischaemic activity: ⊞ indicates the reduction in MRR and force × time (±SEM. 5 subjects) after an approximately constant force tetanus of 1050 impulses at 20 Hz; □ indicates the same (±SEM, 6 subjects) after a declining force tetanus of 6000 impulses at 100 Hz. ■, four myophosphorylase-deficient patients after 1050 impulses at 20 Hz. There is excess slowing in this group.

This is in contrast to normal subjects where recovery can occur during continuing ischaemia (Edwards et al 1981). Interestingly, analysis of the EMG power spectrum in terms of the 'high:low ratio' shows that this declines in a similar way during maximum contractions (Fig. 4) in normal and myophosphorylase-deficient muscle, suggesting that this decline is unlikely to be associated with accumulation of lactate (Lindström et al 1970).

At fatigue in myophosphorylase-deficient muscle ATP is not depleted and phosphoryl creatine concentrations remain higher than normal (Edwards & Wiles 1980a). These results suggest that the immediate cause of force fatigue in such muscle is failure of excitation, although whether this is at the

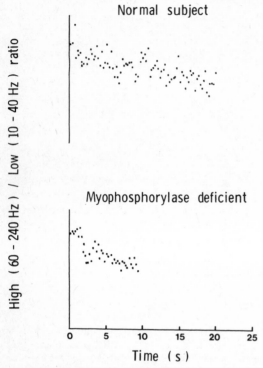

FIG. 4. The EMG power spectrum is shown in terms of the ratio of high and low frequencies (high/low ratio) during maximum voluntary contractions of adductor pollicis in a normal subject (male, 30 yr) and a myophosphorylase-deficient patient (male, 17 yr). The EMG was recorded with surface electrodes. Frequency analysis was done using band widths of 60–240 Hz (high) and 14–40 Hz (low) and an average amplitude for each range was obtained over a period of 200 ms every 240 ms.

neuromuscular junction or at the sarcolemma has not been determined. The failure of recovery of excitation while ischaemia persists and the excess slowing of relaxation referred to above suggest that at least two other energy-dependent systems—the Na^+/K^+-ATPase of the sarcolemma and the Ca^{2+}-activated ATPase in the sarcoplasmic reticulum—may be sensitive to a presumed reduction in ATP flux. This possibility is supported by the abnormal dilatations of the lateral sacs of the triads found in myophosphorylase-deficient muscle when in a state of contracture (Schotland et al 1966) and by the evidence for defective calcium ion reaccumulation by sarcoplasmic reticulum in myophosphorylase-deficient muscle (Gruener et al 1968).

In phosphofructokinase deficiency we find similar premature force fatigue and action potential fade. In the one patient studied, dT/dt in brief maximum contractions was low (0.57 °C min^{-1}). We interpret this as suggesting a more complete block in anaerobic ATP regeneration than is seen in myophos-phorylase deficiency.

Mitochondrial abnormalities. Disorders of mitochondrial function can result in impaired control of the energy supply, as, for example, when the respiratory rate fails to increase in response to increments in phosphate acceptor (respiratory control ratio of unity). Alternatively, the generation of free energy can be reduced, either because of a defect in substrate utilization or transport, or because of a defect in the mechanism of oxidative phosphory-lation. The impaired transport of long-chain fatty acids into the mitochondria for β-oxidation is probably the principal problem in those myopathies that result from a deficiency of carnitine or carnitine palmitoyltransferase (EC 2.3.1.21). Despite the close biochemical relationship of these two deficiency states the clinical syndromes to which they give rise differ considerably, carnitine deficiency presenting as a progressive myopathy with weakness as the main symptom, and deficiency of the enzyme presenting as fatigue, muscle pains and cramps or myoglobinuria late on in exercise or even some hours afterwards. The pathophysiology of these disorders has not been described although it is widely assumed that they result in an inadequate supply of free energy.

Interest has recently centred on a small group of patients with defects in the mitochondrial respiratory chain. Morgan-Hughes et al (1977) described a patient in whom reducible cytochrome *b* was deficient from the electron transport chain of skeletal muscle mitochondria. This patient suffered from pronounced muscle fatiguability and tended to develop lactic acidosis. Later Morgan-Hughes et al (1979) studied two sisters with marked muscle fati-guability, fluctuating muscle weakness and lactic acidosis, and demonstrated an inability to oxidize NAD-linked substrates. This was due to a mitochon-drial lesion at the level of the NADH–coenzyme Q reductase complex (EC 1.6. 99.5). In the former case, and in one of the patients with hypermetabolic myopathy, evidence of failure of excitation–contraction coupling has been found (De Jesus 1974, Morgan-Hughes et al 1977).

We have studied a patient who presented with muscle pains and fatigue on mild exertion, a low maximal oxygen uptake (<1.0 l/min) and profound lactic acidosis, and who also showed impaired oxidation of NAD-linked substrates by skeletal muscle mitochondria.

The energy economy appeared normal, since the rate of metabolic heat production in a maximal contraction was normal (dT/dt, $0.74\,°C\,min^{-1}$), the relaxation rate was normal and the average ATP turnover rate, holding a standard submaximal force to fatigue under ischaemic conditions, was 19.8 μmol ATP s^{-1} mmol^{-1} total creatine (normal; see p 267). As expected, therefore, endurance in the latter test and the ability to sustain force during

electrical stimulation (Fig. 2) was well within normal limits. Muscle performance in ischaemic conditions was therefore quite adequate although, as we
have indicated elsewhere (Edwards & Wiles 1980a), the resting muscle
content of phosphoryl creatine in this patient was well below the normal
range.

We therefore tested this patient in a different experimental system designed to place demands on oxidative metabolism. If normal subjects make
intermittent isometric contractions to 30% MVC (e.g. 1 s contraction every
2 s for 40 s, and then rest for 20 s) they can continue this activity without
discomfort or significant change in the frequency–force relationship of the
muscle for at least 20 minutes. The same pattern of contractions to 60% MVC
for the same total force × time results in minor discomfort and low frequency
fatigue (Edwards et al 1977a), such that the ratio of force produced at 20 Hz to
that in a maximum tetanus is reduced (Fig. 5), although excitation remains
normal. Our patient with a mitochondrial abnormality develops this type of
fatigue at the lower force (30% MVC), in association with considerable

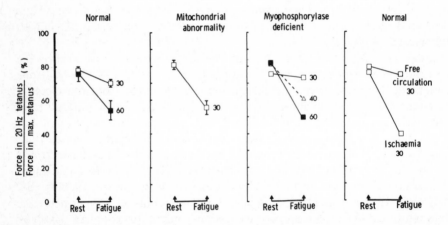

FIG. 5. The frequency–force responses of adductor pollicis or quadriceps measured before (rest)
and 10 min after (fatigue) voluntary isometric contractile activity. Low frequency fatigue was
measured as the decrement in the force produced by a 20 Hz tetanus compared to that in a
maximum (50 or 100 Hz) tetanus. On the left (normal) the mean (±SEM) of this ratio is shown in
6 normal subjects exerting identical amounts of force × time at either 30 or 60% MVC with a free
circulation (see text). The second column (mitochondrial abnormality) shows the mean (±SEM)
of the ratio in 3 tests on one patient contracting at 30% MVC for the same force × time as normal
subjects. The third column (myophosphorylase deficient) shows values in individual tests in one
patient contracting at 30, 40 and 60% MVC (at 30% and 40% MVC the patient could manage the
same force × time as normal subjects but at 60% MVC his muscle fatigued prematurely). The
right column shows the change in ratio when a normal subject contracted under ischaemic
conditions at 30% MVC to fatigue, compared to the change after a matched force × time with a
free circulation. All contractions were made with a free circulation except where shown.

discomfort (Edwards & Wiles 1980a), and is quite unable to complete the test at 60% MVC because of pain and fatigue.

We can assume that a larger proportion of energy for contraction is derived from glycolysis in the 60% MVC series in normal subjects and the 30% MVC series in the patient with abnormal mitochondria than in the 30% MVC series in normal subjects. It is tempting, therefore, to link low frequency fatigue to the development of lactic acidosis, since acidosis considerably increases the amount of calcium required to produce 50% maximum tension in skinned skeletal fibres (Fabiato & Fabiato 1978) and half-maximal myofibrillar ATPase activity (Kentish & Nayler 1979). We therefore studied myophosphorylase-deficient patients in this system and found that they developed increasing degrees of low frequency fatigue at 30%, 40% and 60% MVC (Fig. 5), suggesting that lactate formation *per se* is not the cause.

It seemed possible that this type of fatigue resulted from mechanical factors, perhaps associated with higher muscular tensions. To clarify this we asked normal subjects to make intermittent contractions at 30% MVC to fatigue under ischaemic conditions, and, subsequently, to generate a matched amount of force × time with a free circulation. Low frequency fatigue developed only in the ischaemic series (Fig. 5), suggesting that either ischaemia *per se* or reliance on anaerobic metabolism is the cause of the fatigue. We conclude, therefore, that the patient with a mitochondrial abnormality contracting at 30% MVC with a free circulation behaves like a normal subject contracting at the same relative force in conditions of ischaemia. The reliance on anaerobic energy processes appears to promote low frequency fatigue.

Conclusion

The study of force maintenance in patients with discrete biochemical abnormalities in their muscles can help us to understand the factors normally limiting force production. The results with patients with thyroid disorders indicate the importance of considering the 'economy' of energy utilization as a factor determining fatiguability. Studies in patients with defective glycogenolysis or mitochondrial function suggest that a failure of excitation or excitation–contraction coupling may result from limitation of the energy supply and thus cause fatigue.

Fatigue *in vivo* does not appear to be linked to a major depletion of whole-muscle ATP. Whether specific intracellular sites become critically depleted is unknown. We should consider the possibility that accumulation of metabolites (e.g. H^+), secondary to a biochemical block or as a result of the use of alternative energy-yielding pathways, is the immediate cause of premature fatigue.

Acknowledgements

Our research on energy metabolism in human muscle has been supported by The Wellcome Trust and the Muscular Dystrophy Group of Great Britain. We are grateful to colleagues and others who volunteered to be subjects and to Mr K. Gohil who did the biochemical analyses of needle biopsy muscle samples. The techniques for measuring metabolic heat production were developed as a result of collaboration of R.H.T.E. with Professor D. K. Hill and Mr M. McDonnell at the Royal Postgraduate Medical School, London.

REFERENCES

De Jesus PV 1974 Neuromuscular physiology in Luft's syndrome. Electromyogr Clin Neurophysiol 14:17-27

Di Mauro S, Bonilla E, Lee CP, Schotland DL, Scarpa A, Conn H, Chance B 1976 Luft's disease: further biochemical and ultrastructural studies of skeletal muscle in the second case. J Neurol Sci 27:217-232

Edwards RHT 1978 Physiological analysis of skeletal muscle weakness and fatigue. Clin Sci Mol Med 54:463-470

Edwards RHT, Wiles CM 1980a Energy exchange in human skeletal muscle during isometric contraction. Circ Res, in press

Edwards RHT, Wiles CM 1980b Effect of temperature and ischaemic activity on relaxation rate and heat production in the human adductor pollicis. J Physiol (Lond) 305:83P-84P

Edwards RHT, Harris RC, Hultman E, Kaijser L, Koh D, Nordesjö L-O 1972 Effect of temperature on muscle energy metabolism and endurance during successive isometric contractions, sustained to fatigue, of the quadriceps muscle in man. J Physiol (Lond) 220:335-352

Edwards RHT, Hill DK, Jones DA 1975 Heat production and chemical changes during isometric contractions of the human quadriceps muscle. J Physiol (Lond) 251:303-315

Edwards RHT, Hill DK, Jones DA, Merton PA 1977a Fatigue of long duration in human skeletal muscle after exercise. J Physiol (Lond) 272:769-778

Edwards RHT, Young A, Hosking GP, Jones DA 1977b Human skeletal muscle function: description of tests and normal values. Clin Sci Mol Med 52:283-290

Edwards RHT, Young A, Wiles CM 1980 Needle biopsy of skeletal muscle in the diagnosis of myopathy and the clinical study of muscle function and repair. N Engl J Med 302:261-271

Edwards RHT, Wiles CM, Gohil K, Krywawych S, Jones DA 1981 Energy metabolism in human myopathy. In: Schotland DL (ed) Disorders of the motor unit. Houghton Mifflin, Boston, in press

Fabiato A, Fabiato F 1978 Effects of pH on the myofilaments and the sarcoplasmic reticulum of skinned cells from cardiac and skeletal muscles. J Physiol (Lond) 276:233-255

Gruener R, McArdle B, Ryman BE, Weller RO 1968 Contracture of phosphorylase deficient muscle. J Neurol Neurosurg Psychiatr 31:268-283

Haydar NA, Conn H L, Afifi A, Wakid N, Ballas S, Fawaz K 1971 Severe hypermetabolism with primary abnormality of skeletal muscle mitochondria. Ann Intern Med 74:548-558

Kentish JC, Nayler WG 1979 The influence of pH on the Ca^{2+}-regulated ATPase of cardiac and white skeletal myofibrils. J Mol Cell Cardiol 11:611-617

Lambert EH, Underdahl LO, Beckett S, Mederos LO 1951 A study of the ankle jerk in myxedema. J Clin Endocrinol 11:1186-1205

Lindström L, Magnusson R, Petersén I 1970 Muscular fatigue and action potential conduction velocity changes studied with frequency analysis of EMG signals. Electromyography 4:341-356

Luft R, Ikkos D, Palmieri G, Ernster L, Afzelius B 1962 A case of severe hypermetabolism of non-thyroid origin with a defect in the maintenance of mitochondrial respiratory control—a correlated clinical, biochemical and morphological study. J Clin Invest 41:1776-1804

McArdle B 1951 Myopathy due to a defect in glycogen breakdown. Clin Sci (Oxf) 10:13-35

Morgan-Hughes JA, Darveniza P, Kahn S N, Landon DN, Sherratt RM, Land JM, Clark JP 1977 A mitochondrial myopathy characterised by a deficiency in reducible cytochrome b. Brain 100:617-640

Morgan-Hughes JA, Darveniza P, Landon DN, Land JM, Clark JB 1979 A mitochondrial myopathy with a deficiency of respiratory chain NADH-CoQ reductase activity. J Neurol Sci 43:27-46

Schotland DL, Spiro D, Carmel P, Rowland LP 1966 Ultrastructural studies of muscle in McArdle's disease (deficiency of muscle phosphorylase). J Neuropathol Exp Neurol 25:146-147

van Wijngaarden GK, Bethlem J, Meijer AEFH, Hülsmann WC, Feltkamp CA 1967 Skeletal muscle disease with abnormal mitochondria. Brain 90:577-592

Wiles CM, Edwards RHT 1979 Energy turnover and fatigue in human muscle. Clin Sci (Oxf) 57:1-2P

Wiles CM, Young A, Jones DA, Edwards RHT 1979 Muscle relaxation rate, fibre type composition and energy turnover in hyper- and hypothyroid patients. Clin Sci (Oxf) 57:375-384

Wiles CM 1980 The determinants of relaxation rate of human muscle in vivo. PhD thesis, University of London

DISCUSSION

Wilkie: You described a patient with a mitochondrial deficiency who could not use the oxidative pathway and didn't take in much oxygen (maximum oxygen intake <1.0 l/min). What did he do with the oxygen he did take in?

Wiles: The block in oxidative metabolism was not complete, in that he was able to oxidize flavoprotein-linked substrates such as succinate through the electron transport chain. The amount of oxygen he could utilize appeared appropriate for the low level of exercise he was able to undertake.

Morgan-Hughes: I was interested in the mechanisms of fatigue in the two patients with different types of metabolic defect. As I understand it, in ischaemic myophosphorylase-deficient muscle there appears to be a failure of excitation which persists until the blood supply is restored. In muscle which is already functioning anaerobically, because of a defect in oxidative metabolism, fatigue is associated with failure of excitation–contraction coupling. We have found a similar defect in other patients with mitochondrial lesions (Morgan-Hughes et al 1977, 1979). Do these findings imply differences in the way in which available energy is used in these two types of defect? Professor

Wilkie spoke earlier about ATP being freely available for all energy-requiring processes. Why, then, should the mechanisms of fatigue be so different in these two conditions?

Wilkie: Isn't it simply that in the patient in whom mitochondrial oxidative phosphorylation is defective, ATP can be regenerated only through glycolysis? We have agreed that in isolated frog muscles and in human exercising muscles the ATP level never falls below about 50% of its normal level. So one wouldn't expect it to fall in such patients, because they would soon be dead if the ATP concentration did fall. It seems reasonable that the patient with the mitochondrial deficiency had to regenerate all his ATP from ADP by glycolysis.

Edwards: The question is whether the alterations in excitation or in excitation–contraction coupling in these patients can be explained on the basis of alterations in energy supply.

Jones: The situation is that one patient (with the mitochondrial defect) is functioning as if he were a normal person working anaerobically and the other (with myophosphorylase deficiency) is functioning as if he were a normal person but poisoned with iodoacetate.

Wilkie: So far in the symposium we have taken fatigue through the nervous system as far as the action potential (and the rapid recovery process is again a matter of action potentials and ions). We have also discussed the more biochemical approach, looking at actomyosin ATPase and so forth, but there is still a vast gap. Calcium has hardly been mentioned. There is the evidence that concentrations of metabolites in some way influence excitation (Lüttgau 1965, Dawson et al 1978, 1980a, b), but the details of how and where remain mysterious. This is what has to be brought together.

Pugh: Do these enzyme deficiencies affect other tissues besides skeletal muscle?

Wiles: That depends on the particular enzyme. In myophosphorylase-deficient patients the abnormality is probably confined to skeletal muscle. In phosphofructokinase deficiency the red blood cells are affected as well. There is a suspicion that liver function is affected in the patient with a mitochondrial defect, in whom the clearance of lactate from the circulation is very slow.

Macklem: In some patients with metabolic abnormalities of muscle the diaphragm is spared, I think?

Morgan-Hughes: The diaphragm may be selectively involved in acid maltase deficiency (Newsom Davis et al 1976) but I am not aware that it is ever spared in McArdle's disease. Myophosphorylase deficiency seems to affect muscle function in different ways. It sometimes gives rise to progressive weakness rather than exercise intolerance (Engel et al 1963). DiMauro & Hartlage (1978) recently described an infant girl with rapidly progressive weakness which began soon after birth and led to fatal respiratory failure at the age of 13 weeks. Another puzzling feature in McArdle's disease is that symptoms do not usually appear until late childhood or early adolescence, yet the enzyme deficiency has presumably been present since birth. Indeed in one patient (Kost & Verity 1980), muscle cramp and myoglobinuria first appeared at the age of 60. It is difficult to explain such clinical heterogeneity on the basis of any single biochemical mechanism.

In general, I find myself left with the problem of trying to relate the laboratory findings to the patient's symptoms. Do the various mechanisms of fatigue presented here apply to everyday life, or are we looking at extreme situations? Perhaps something much more subtle is happening at an earlier stage. Fatigue in a clinical context is more an overwhelming sense of weariness than a loss of force. Patients who complain of fatigue do not suddenly fall down in the street. Perhaps clinical fatigue is due to a combination of factors rather than to any single event.

Edwards: I would hope that most of the analysis we have heard about in this meeting is at least a starting point for analysing fatigue in patients and helping to decide whether their fatigue is in the mind or in the muscle. Dr Wiles has shown that a subtle interrelationship between energy exchange and electrical phenomena may be responsible for fatigue in some patients.

Stephens: As I understand you, Dr Wiles, if you cool the adductor pollicis muscles of your normal subjects they can maintain the same force for longer. I would have expected muscle to work best at normal body temperature.

Wiles: It does seem that in daily life a compromise must be reached between contractile speed and energy economy. One's fingers are less nimble at 27 °C than 37 °C (intramuscular temperature) because of the slowed contractile properties of the muscles—at least in part. On the other hand, the muscles are more economical in their use of energy and are less fatiguable. Like the tortoise and the hare, one pays for speed with increased energy cost and reduced endurance. It is a corollary of this argument that cooled and hypothyroid muscle should sustain trains of action potentials better than normal warm muscle, if indeed they are more resistant to fatigue.

Edwards: This brings us back to the membrane again, as something which is influencing the energy economy. Thus in hypothyroidism there is a reduced overall metabolic rate, yet we have preservation of action potentials and less tendency to become fatigued.

Wiles: In all the normal subjects and patients we have examined using sub-maximal (20 Hz) trains of stimuli to obtain the optimum force × time per impulse, we find that force fatigue is accompanied by fading of the action potential—at least in ischaemic conditions. We cannot therefore argue that force fatigue is due to a reduced energy supply, unless this acts through the excitation/activation mechanism, or both fail in parallel but for different reasons.

Dawson: You say that metabolic disease might cause fatigue through a change in the activation of contraction in some patients. Would you also say that when you or I are fatigued, with a decline in the maximum force we can produce, that may also be due to a change in the metabolic state of the muscle through an effect on the activation of contraction?

Wiles: Yes.

Dawson: Dr Roussos, you also felt that fatigue in respiratory muscles was due to a change in their metabolic state, perhaps an increase in the concentration of lactic acid. However, you said that force development in fatigued muscles is increased by giving aminophylline, probably as a result of this drug's effect on calcium release from the sarcoplasmic reticulum. Taking these two lines of evidence together, would you deduce that in your experiments, fatigue is due to metabolic changes which in turn affect the activation of contraction?

Roussos: We don't know the mechanism of diaphragmatic fatigue in our experiments. We know that blood and muscle lactate concentrations increased and pH decreased. We also know that low pH impairs the binding of calcium with troponin and perhaps sequesters calcium in the sarcoplasmic reticulum. Aminophylline is known to improve the release of calcium from the sarcoplasmic reticulum, so *if* the impairment of the diaphragm in our experiments lies in the excitation–contraction coupling process, the release of calcium might improve contraction. But this is all hypothetical!

Dawson: Yes, but both you and Dr Wiles are going in the same direction.

Saltin: Have you any information on potassium release in isometric contractions in your patients, especially in the myophosphorylase-deficient patients as compared to the patient with abnormal mitochondria? Hník et al (1976) showed that potassium release is mainly due to the net loss of potassium in the activation of the fibre, whereas I think that glycolysis may play a role.

Wiles: I don't know about potassium release in our patients.

Edwards: Barcroft showed an increased release of phosphate in a patient with myophosphorylase deficiency (Barcroft et al 1971). I have no information on potassium.

Campbell: What firing rate would you expect in the motor neurons of these patients, in a maximal effort, or for that matter in the normal subject with low frequency fatigue due to ischaemia?

Wiles: I would expect it to be similar to a normal resting subject. If the contractile speed of the muscle is normal, the firing rate required to produce a fully fused tetanus or maximum voluntary contraction should be unchanged.

Jones: One change in a muscle that has been working very hard is that the amount of calcium released from the sarcoplasmic reticulum appears to be reduced. This may be happening during a contraction; it certainly persists for a long while afterwards (up to 24 hours) when the levels of metabolites are back to normal and so is the action potential (Edwards et al 1977).

Wilkie: What is the direct evidence for that? I don't know any experiments using calcium detectors that show this reduced release of calcium.

Jones: The evidence is indirect. A reduced force at low frequencies and normal force at high frequencies with normal metabolite levels and electrical activity implies either a reduced release of calcium, or reduced sensitivity of the regulating proteins for calcium.

Wilkie: There are so many things we don't know about the calcium-activating system that a negative argument isn't very strong. Proteins like parvalbumin and calmodulin, which buffer calcium, are found in large quantities in various types of muscle; their regulatory function remains largely a matter for speculation. So you can't say that it *cannot* be anything apart from calcium release that is responsible for variations in response.

Jones: I think we can say that there are long-lasting changes somewhere between the release of activator and its action on the actomyosin. The slow time course of recovery suggests to us repair of damaged membranes, probably involving protein synthesis, rather than a change in the conformation or phosphorylation of a regulatory protein.

Edwards: We see defects in activation in human muscle in circumstances—some predictable and some rather surprising—specifically related to energy exchange.

Wiles: We should be able to get at this problem by looking more closely at the time course of recovery of the defect in excitation–contraction coupling which appears to follow contractions that utilize anaerobic energy resources, and seeing what metabolic event it could relate to—perhaps the clearance of the proton load generated during a contraction. The time course of this is not known. We know the time course of lactate clearance from normal muscle but that may differ from proton clearance (N. L. Jones 1980).

Edwards: Lars Hermansen showed us something of the time course of changes in pH in muscle samples (see p 75–82).

Hultman: Have you done dynamic exercise tests of the patients with myophosphorylase deficiency? They should have a decreased capacity for hard muscular work.

Wiles: We haven't done this systematically. One might anticipate that their aerobic capacity would be normal, since if you can get such patients over the first hump of exercise, they usually get their 'second wind'. Once their oxidative metabolism is fully switched on I don't see why it shouldn't be normal.

Saltin: Patients with myophosphorylase deficiency can't reach high work outputs.

Wiles: Although our patients with myophosphorylase deficiency are not weak, some do develop a late-onset myopathy. This may be a factor in reducing their working capacity, quite apart from any problems they may have in utilizing anaerobic energy resources.

Saltin: It is a question of how much substrate is available for the muscles. In all types of peak performance in man glycogen has to be utilized. If the work output is limited by the uptake of free fatty acids and glucose from the bloodstream and the usage of triglyceride in the muscle, the work capacity is limited.

REFERENCES

Barcroft H, Foley TH, McSwiney RR 1971 Experiments on the liberation of phosphate from the muscles of the human forearm during vigorous exercise and on the action of sodium phosphate on forearm muscle blood vessels. J Physiol (Lond) 213:411-420

Dawson MJ, Gadian DG, Wilkie DR 1978 Muscular fatigue investigated by phosphorus nuclear magnetic resonance. Nature (Lond) 274:861-866

Dawson MJ, Gadian DG, Wilkie DR 1980a Mechanical relaxation rate and metabolism studied in fatiguing muscle by phosphorus nuclear magnetic resonance. J Physiol (Lond) 299:465-484

Dawson MJ, Gadian DG, Wilkie DR 1980b Studies of the biochemistry of contracting and relaxing muscle by the use of ^{31}P n.m.r. in conjunction with other techniques. Proc R Soc Lond B Biol Sci 289:445–455

Di Mauro S, Hartlage PL 1978 Fatal infantile form of muscle phosphorylase deficiency. Neurology 28:1124-1129

Edwards RHT, Hill DK, Jones DA, Merton PA 1977 Fatigue of long duration in human skeletal muscle after exercise. J Physiol (Lond) 272:769-778

Engel WK, Eyerman IL, Williams HL 1963 Late onset type of skeletal muscle phosphorylase deficiency. A new familial variety with completely and partially affected subjects. N Engl J Med 268:135-137

Hník P, Holas M, Krekule J et al 1976 Work-induced potassium changes in skeletal muscle and effluent venous blood assessed by liquid ion-exchanger micro-electrodes. Pflügers Arch Eur J Physiol 362:85-94

Jones NL 1980 Hydrogen ion balance during exercise. Editorial review. Clin Sci (Oxf) 59:85-91

Kost GJ, Verity MA 1980 A new variant of late-onset myophosphorylase deficiency. Muscle Nerve 3:195-201

Lüttgau HC 1965 The effect of metabolic inhibitors on the fatigue of the action potential in single muscle fibres. J Physiol (Lond) 178:45-67

Morgan-Hughes JA, Darveniza P, Kahn SN et al 1977 A mitochondrial myopathy characterized by a deficiency in reducible cytochrome b. Brain 100:617-640

Morgan-Hughes JA, Darveniza P, Landon DN, Land JM, Clark B 1979 A mitochondrial myopathy with a deficiency of respiratory chain NADH–CoQ reductase activity. J Neurol Sci 43:27-46

Newsom Davis J, Goldman M, Loh L, Casson M 1976 Diaphragm function and alveolar hypoventilation. Q J Med 45:87-100

GENERAL DISCUSSION

The perception of fatigue

Edwards: In this final discussion we should consider the perception of fatigue and decide whether the definition of fatigue proposed initially for the purpose of this symposium is an adequate one. Perhaps John Morgan-Hughes could start us off? He pointed out earlier that to the practising neurologist, the physiologist's definition was not of practical use.

Morgan-Hughes: The question arises of whether the electrical and mechanical changes that have been described are responsible for fatigue in the clinical sense. I do not know how often I fire my motor units at 120 Hz or even at 80 Hz. Perhaps these changes are relevant to the trained athlete competing in an Olympic stadium, but do they apply in the out-patient clinic? When trying to analyse the symptom of fatigue one runs into semantic difficulties. Patients will often say that they 'have run out of energy' but this does not mean that they lack energy in the biochemical sense. Perhaps clinical fatigue is due to the summation of a number of events which precede the changes recorded in the laboratory.

Campbell: I tried to indicate earlier that I don't think the sense of fatigue is just a sense of weakness, or at least weakness as produced by neuromuscular block (p 171). I suspect that the sense of muscular fatigue has three ingredients: firstly, a greater sense of the effort required to accomplish a task; secondly, weakness, but this is not the most striking feature; and thirdly, sensations in the fatigued muscle that are present even if it is not being used. I do believe that there is a sense of effort that can be separated from the sense of developed force, and a component of the fatigue may be having to make a greater effort to develop the same force in the muscle, joint or tendon.

Edwards: Pat Merton at a previous Ciba Foundation symposium gave an account of his view of the neurophysiology of the sense of effort (Merton 1970). Do you want to add anything here, Pat?

Merton: I agree with everything Dr Campbell says. Fatigue is not the same as weakness produced by curare, and the feeling of weakness is not the same as the feeling of fatigue. And there is a sense of effort, which is distinguishable from a sense of force actually developed.

Edwards: The experiments are not really satisfactory in separating them, of course, but there have been good attempts to do just that. In your 1970 paper you quoted the arguments of Helmholz in relation to eye movements. The question is the extent to which one has a sense of effort because one is initiating a voluntary act, and to what extent it is a function of the fact that one has achieved (or not achieved) that act.

Merton: An example suggested directly by Sherrington (1900) is that if you make minute movements of a slide under a microscope using your fingers, you don't know whether you have succeeded unless you watch in the microscope. You can *will* yourself to make much smaller movements than you can perceive by position sense in the fingers. This illustrates that you have a sense of effort which is separate from the perceived results of volition.

Pugh: This sense of effort can only mean that you have to recruit more groups of muscle fibres to produce a given result. You see this clearly in runners, and you can measure an increase in oxygen intake at a given speed because of this. So the sense of increased effort is not wholly subjective.

Campbell: Nobody seems to object to my third category to which the term 'fatigue' is applied—that if you are doing nothing after you have done something fatiguing, you feel differently. I don't know whether the sensation arises in muscles, joints or tendons, but it is part of the total complaint that patients have, and also normal subjects after exhausting exercise.

Edwards: Brodal (1973) made observations on himself after suffering a stroke and described with great precision all his sensations, including the period of weakness. At the time it was a great effort to do simple tasks and he became fatigued very quickly.

Morgan-Hughes: This is a common symptom in patients with hemiplegia.

Edwards: Patients with motor neuron disease experience this.

Morgan-Hughes: I agree. Patients with demyelinating disease also complain of intense fatigue, particularly during a relapse, yet muscle power when tested clinically may be normal.

Edwards: It seems to be possible to distinguish whether the problem is in the machine or the driver of the machine. If we are dealing now with someone in whom the problem of fatigue is in the central nervous system, can we think of mechanisms by which it has been brought about? The fact that such a person can generate normal power in a brief contraction suggests that at least for short times he or she can maintain the necessary high firing frequency or motor unit recruitment. The problem is then one of reflex potentiation, or whatever it is that keeps up the firing frequency.

Morgan-Hughes: Yes. It has been shown experimentally that transmission of impulses through a demyelinated zone begins to fail at high stimulus frequencies (McDonald & Sears 1970).

Edwards: That is interesting, because it occurred to me, thinking of high frequency fatigue in myasthenia gravis, that such patients, even if their force on formal testing was good, might fall on tripping because they cannot make the rapid contractions necessary to correct their balance, because of high frequency fatigue. It might be a simple test to see what is the fastest contraction that can be made. Can patients with demyelinating diseases make ballistic movements?

Morgan-Hughes: I don't know.

Edwards: That would be a test of how quickly you can set the machine going. You can then separate that from the problem of keeping the machine at work.

Wiles: The patients who have difficulty with sudden or ballistic contractions are those with basal ganglia disorders (Hallett & Khoshbin 1980). Normal subjects and these patients were asked to make ballistic contractions of different angular distances. The normal subjects made ballistic contractions over different angular distances in the same time. They do this either by recruiting more motor units or by using a higher firing frequency to traverse that distance. The EMG activity occurs over the same period of time but is more intense for the longer movement. Patients with early Parkinson's disease were unable to do this. They took longer to produce bigger angular deviations and the amplitude of the EMG failed to increase. In basal ganglia disturbances, therefore, the ability to energize specific groups of muscles at high speeds, as in a ballistic contraction, is the major problem.

Merton: How relevant is this to fatigue? Parkinsonian patients with akinesis don't complain of fatigue, do they? It isn't that they can't do up their buttons because they feel fatigued; they just can't do it.

Edwards: We got onto Parkinson's disease from Brodal's observations on himself and the fact that fatiguability of a central type seems to be a feature of demyelinating disease.

Campbell: I do think it is possible to separate the sense of effort from the performance. You may not know what effort you have made just in terms of what you accomplish. This belief, even if validated, doesn't necessarily clarify fatigue if you define fatigue in terms of motor failure only, because the sense of effort drops out of the definition unless that sense of effort is also accompanied by loss of motor performance. I am sure that fatigue in normal experience includes a component of a discrepancy between the effort I have to make and what I accomplish, *even* when I can accomplish 100%.

Merton: Is the nervous syndrome with the greatest dissociation of effort and result perhaps cerebellar ataxia, where patients try to do something and something quite different happens? But their complaint is not of fatigue. They have a gross mismatch between their effort and their achievement.

Lippold: Nobody has mentioned yet that the sensations associated with fatigue are unpleasant, and possibly directed towards stopping the contraction producing the fatigue!

Campbell: Yes. And this again is different from being curarized, which, provided the consequences of being paralysed are avoided, is not at all unpleasant!

Stephens: On the mismatch between effort and what you finally produce, if you ask a subject to maintain a steady contraction at a sub-maximal level, record the electrical activity in the muscle, and then wait, after a while the electrical activity starts to rise as the subject has to recruit extra units to compensate for those that are already fatigued. It is at that moment that the subjects say that they are starting to fatigue. They know they are starting to fatigue, the moment they have to begin putting more effort into their contraction. You can tell they are having to do that because new motor units are recruited and muscle electrical activity starts to rise.

Electricity or chemistry?

Edwards: May we now consider Pat Merton's earlier contention (p 152) that muscle fatigue has everything to do with electricity and nothing to do with chemistry? We can do this best in the form of a debate. I shall ask Pat Merton to propose this motion, to be seconded by Brenda Ritchie. To oppose, I suggest Eric Hultman and Douglas Wilkie.

Merton: There have been suggestions made already during the symposium that in some circumstances one can exhaust the excitation–contraction coupling system long before one exhausts the system of sliding filaments and ATP. With stimulation at, say, 60/s there is no reason to suppose that there is any early failure of neuromuscular transmission, so the muscle action potential continues but it fails to cause the muscle to contract fully. That might be due to using up the 'coupling agent' in the excitation–contraction coupling mechanism. With a skilful regime of stimulation you can spin out the store of coupling agent and so get a longer contraction out of the muscle, with a correspondingly greater use, possibly a complete use, of the energy-rich compounds in muscle that directly fuel the contraction. That is the thesis: that what you get out of the muscle depends critically on how you stimulate it—put in an epigrammatic form for the purpose of debate.

The evidence for this contention is in Fig. 4 (p 152). Stimulation at 60/s, suitably tapered off to 20/s, gives $2\frac{1}{4}$ times as much tension-time as a tetanus at 60/s containing the same number of impulses. The particular tapered regime of stimulation used appeared to be optimal, since it was found by trial and error that any faster or slower tapering gave less contraction. Why evolutionary pressures have produced this particular performance as optimal is obscure. It might have turned out that the contraction could be held at its initial maximal value until coupling agent and energy stores were exhausted, when it would collapse rapidly—like an ideally designed car that would run at full performance until everything wore out simultaneously. But that is not what happens.

To sum up by putting the argument in a slightly different form: Fig. 4 (p 152) shows that with the frequencies used (which are physiological) the tension depends only on the number of motor impulses received and not on their distribution in time. As an epigram: electricity governs. It is argued that chemical changes cannot govern, because they are most unlikely to be identical in two contractions differing in duration by a factor of $2\frac{1}{4}$.

Wilkie: Would it be helpful if we mentioned what we would regard as experimental methods for distinguishing between hypotheses? In particular, there is an experiment that hasn't been done that would be relevant here. Presumably if, in the two curves where force is plotted against number of stimuli, there were parallel chemical determinations that showed that when tension had fallen nearly to zero in both cases the same amount of the same chemical reactions had occurred, you would concede that both 'electricity' and 'chemistry' play a part together?

Merton: I have no evidence about chemical changes except circumstantial evidence.

Edwards: It is one thing to say that force is related to the number of impulses; that is an observation, but if you try to tie this to a mechanism—and here we can extend the idea of 'chemistry' into the chemistry of excitation— you have to explain this in terms of the pumping of ions with totally different time courses. How is it that you appear to have delayed the restitution of sodium and potassium exchange across the membrane during the period when you were stimulating at low frequency?

Merton: Fig. 4 (p 152) showed a very good coincidence between a contraction that started at 60/s and fell to 20/s and another which was kept at 60/s throughout, when plotted against number of impulses. Fig. 1 shows the

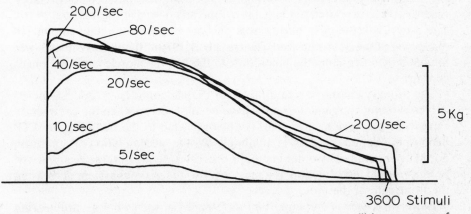

FIG. 1. (Merton). Tracings of records of the force produced by adductor pollicis over a range of frequencies of stimulation from 5/s to 200/s, plotted against the number of stimuli delivered (intended to be 3600 in each case). The records were made by altering the speed of the recording paper in proportion to the rate of stimulation. The fastest record lasted 18 s and the slowest 12 min. (C. D. Marsden, J. C. Meadows & P. A. Merton, unpublished.)

results for a number of fixed-frequency tetani plotted against number of stimuli, 3600 in each case. The general similarity of the curves is striking. That is the circumstantial evidence. The curves from 20 up to 200/s are not much different. It may be that the chemistry is all the same, but I think that is unlikely over that range of frequencies, which gives durations of contraction differing by a factor of 10 or more.

Wilkie: Here we have a testable hypothesis. It is perfectly possible to do the parallel chemical determinations.

Edwards: Eric Hultman will now answer Pat Merton from the other side of the floor.

Hultman: In skeletal muscle we have a series of stores of energy substrates that can be used to fulfil the demands elicited by the nervous impulse. There are energy-rich phosphagens, glycogen and fat. Substrates from the circulating blood such as glucose, free fatty acids and ketone bodies can also be used.

During dynamic exercise the glycogen store can be utilized practically completely, and the point of fatigue will coincide with the emptying of the glycogen store. You are tired for biochemical reasons.

The same fatigue can be experienced with decreasing levels of blood glucose. The fatigue in this situation is experienced primarily in the central nervous system.

The phosphocreatine store can also be utilized completely during hard voluntary exercise, dynamic or isometric. The length of contraction in this situation will also be determined by the size of the energy store utilized. It was shown that if repeated isometric contractions were performed to exhaustion with varying rest intervals (5–40 s) between the contractions, the performance time was directly related to the length of the rest period (Edwards et al 1971). The length of the rest period was shown to determine the amount of phosphocreatine resynthesized (Harris et al 1976). If this resynthesis was inhibited by obstructing the blood flow to the leg, the muscle could no longer contract.

The primary energy source utilized for muscle contraction, ATP, is never depleted. Only marginal decreases have been observed when the contraction capacity is near exhaustion. This is obviously due to the fact that the ATP dephosphorylation process is inhibited by the accumulation of products normally eliminated by other energetic processes. Some of these accumulated products could also function as a messenger to the nervous system decreasing the firing rate of the motor nerve.

Edwards: There is one important difference between what we are hearing from Eric Hultman and Pat Merton. Eric Hultman is saying that when you can't continue to exercise, the muscle will still contract and the subject is able to stand up and walk away, but he can't pedal a bike any longer. It is a question of defining fatigue as failure to continue at a particular power output, as opposed to generation of force. Eric Hultman is talking about maintaining a particular power output.

Wilkie: The definition of fatigue as a decline in isometric force has been put forward as a practical, measurable definition; but I have never accepted the need for a single definition. There are various things that alter in different ways. The function of muscles is not merely to exert static forces but also to do external work, and one should consider both together. The effect of temperature, for example, is negligible on isometric force development, over a reasonable range, whereas at lower temperatures the maximum mechanical power production is much reduced, so one must think of that as well. Here I am sticking strictly to objective, measurable quantities, where you ask the patient who says he is tired, 'what *can't* you do?'

Edwards: Will Dr Ritchie say something in reply, from the other side?

Bigland-Ritchie: I agree that in many situations a reduced work capacity may be a more relevant definition of fatigue than one limited to force alone. However, in experiments such as ours where only isometric contractions have been studied the force–time integral is the corresponding parameter. In that context I am not sure how meaningful is the question we are debating. Obviously fatigue cannot be 'all in the electricity', or 'all in the chemistry': both are required. If the activation is not delivered to the muscle at a rate which corresponds to the changing propagation capacity of the electrical transmission system, or the changing contractile and metabolic requirements of the muscle, the available metabolic energy cannot be utilized in an optimal manner. However, if the activation rate is adjusted appropriately by changes in motor neuron firing frequencies (or even conceivably by appropriate rates of neuromuscular block), we have not found evidence that reduced activation limits the contractile process. If that is so, the loss of force must be due to limitations of either excitation–contraction coupling or the rate of metabolic energy supply, and chemistry wins: but not unless the motor neurons know what they are doing!

Wilkie: I agree with Pat Merton about many things, in fact! I agree that if muscles were 'better designed' they would maintain maximum force and power output until they couldn't go on any longer, because there are no biological situations that I can think of where it would not be an advantage to have the choice of doing so. After all, you could always slow down voluntarily. The motor car analogy proposed by Dr Merton is an excellent one here.

It is striking that throughout this symposium, apart from a comment by Lars Hermansen on glycerinated (or skinned) muscle fibres, no one has spoken about the 'works'—the actomyosin system and its properties and how they might be influenced by chemical or other changes. There are the two extreme hypotheses about fatigue, that there is something wrong with energy transduction processes in the muscle fibrils, or that something is wrong with activation, but they are not mutually exclusive. In fact, we have reasons to think that both operate. We have considered activation quite thoroughly; let us now look at events in the muscle fibril where ATP is hydrolysed to ADP, inorganic phosphate (P_i) and protons, the precise stoichiometry and the charge numbers depending on the pH of the reaction:

$$MgATP + H_2O \rightarrow MgADP + P_i + H^+$$

The reaction is catalysed by actomyosin. In the fatigued muscle the ATP concentration doesn't alter much, but the concentrations of all the products are increased considerably. The ADP concentration increases at least 15-fold, in our experiments in frog muscle. P_i goes up more than 10-fold; H^+ concentration also goes up almost as much. The kinetic scheme for the hydrolysis of ATP by actomyosin is complicated, but in general one would expect that when the concentration of all three products goes up many times, the reaction would be slowed down. So the most obvious and testable hypothesis about the decline in force generation and power production is simply that the rate of splitting of ATP by the enzyme is reduced because of product accumulation. The irritating thing about this situation is that it suggests a perfectly feasible experiment. Quite a number of biochemists have *in vitro* actomyosin systems in their laboratories and could do the experiment if they wished. Joan Dawson and I have long been urging them on. Unfortunately, the notion of changing three variables simultaneously goes against the grain with biochemists, so none of them has actually done the experiment! It could also be done with chemically or mechanically skinned fibres, to see whether their work production and force production are reduced. The result would illuminate the question of whether it is the performance of the contractile mechanism that is being reduced by product accumulation.

As for the question of why the ATP level doesn't fall, this probably indicates a control point or control design, although, as Pat Merton mentioned, perhaps this control is so designed (and this would certainly be sensible) that when the muscle is approaching the dangerous point of depletion of ATP, the 'switching' on of the actomyosin system also declines so that you can't run yourself into rigor mortis!

There is another (and testable) possibility. *In vitro*, actomyosin can hydrolyse ATP down to very low levels, so there is something intrinsically puzzling about the fact that *in vivo* the level never falls below about 3 mM ATP. However, since the experiment testing the hypothetical role of product accumulation has not been done, I don't know whether, in the presence of products, actomyosin continues to hydrolyse ATP down to these low levels. Perhaps it would not, in spite of the fact that it remains thermodynamically highly spontaneous. I am still trying to persuade someone to do that experiment also.

Assessment of design of course is subjective. In my view it would have been much better to have muscles which function at full strength until you run out of substrate, without your slowing down. After all, it is slowing down that makes you into somebody else's dinner! Likewise, muscles are ill-designed in the sense that you have to use up chemical fuel merely to exert a static force, which is absurd. An elastic band will hold a force for ever without using up any chemical fuel. There are a few muscles (some of those that hold the shells of bivalves closed) that do maintain a force for long periods with almost no consumption of chemical fuel. But there seems to be something intrinsic in the sliding filament design that requires trade-off between the efficiency with which work can be produced and the economy with which force is maintained. Partly, that is why in our design it is good that we have different sorts of muscle fibres, each apparently adapted to be good at one of the two jobs—maintaining force at low cost and doing work at low cost.

The third point about design is why the efficiency of muscles is so low. Taken overall from oxygen it is 20–25%, so 75–80% of the free energy that might potentially be converted into work is being wasted, in spite of great evolutionary pressure. You can show that having more efficient muscles would be a great advantage to an animal in catching its dinner or in avoiding being caught for dinner (see Wilkie 1974). You could make a much more efficient animal altogether: it wouldn't need so big a heart if the heart muscle were more efficient and the skeletal muscles wouldn't need so big a blood supply. So there must be some fundamental limitation that makes muscle inefficient. We don't even know how much of the inefficiency is in the transduction of energy from chemical free energy into work and force, and how much is lost in the purely chemical factory processes by which ATP is regenerated from ADP by glycolysis and oxidative phosphorylation. No work is involved here, but free energy is degraded into heat.

This all provides substantial reasons for thinking that the contractile mechanism should not be forgotten and is intended to counterbalance what I consider to be too much stress on the switches!

Merton: That is absolutely right—chemistry *ought* to govern; I suspect it doesn't when a physiologist takes hold of a stimulator and stimulates a muscle too fast. If I may tease Dr Dawson and Professor Wilkie, probably 50/s is about the worst rate. It's not quite fast enough to give the maximum contraction and not slow enough to prevent fatigue!

Wilkie: It depends very much on the preparation.

Bigland-Ritchie: That was my point too.

Edwards: While we are arguing from an evolutionary point of view it is worth remembering that evolution has not affected muscle only, but also muscle–bone mechanical properties. In the book *Scale effects in animal locomotion* a distinction is made between flapping and soaring birds and the different requirements for muscle for completely different patterns of flight (see Pedley 1977). One is therefore dealing with quantal moves in evolution which don't follow a gradual, continuous function for a single structure such as muscle. So far as avoiding being caught for dinner goes, 'a miss is as good as a mile', and the evolutionary factor that may be most important is the rate of generation of force, instantaneous power output, rather than the energy economy of the muscle. The needs of survival may have to take priority over considerations of economy.

Wilkie: It depends on the habit of life of the particular species. Some species, like rabbits, survive by sprinting to their holes. Hares survive by being able to run for long distances. According to different circumstances in different species, different features are optimized.

Merton: It seems disappointing for that optimistic view of evolution that the quadriceps muscle, the diaphragm and the adductor pollicis all have the same relationship between frequency and force (see Moxham et al, this volume, Fig. 1, p 200). That wasn't what was expected!

Stephens: May I put a question to the chemists? I suspect that there is no single mechanism of fatigue, and that sometimes the switches fail, sometimes something goes wrong with the enzyme systems, and so on. We would learn a lot if we could find the special circumstances in which each fails. Leaving that aside, can we learn anything about mechanisms of fatigue by asking why the different types of muscle fibre fatigue along different force profiles? Histochemically these muscle fibre types look different, and they fatigue at different rates. Would you, Dr Merton, argue that the histochemistry matches the 'switch', which is different in different muscle fibre types?

Merton: I would hope so. I really hope that in ordinary life it is chemistry that governs and that these peculiarities show up only when you stimulate the muscle machine with 100 shocks a second or in some other excessive manner, which it is not designed for.

Wilkie: On this question of whether the switch fails first, Eric Hultman has told us that when the fuel runs out you can't go on. I want to put the question the other way round: do you ever find a subject who can't go on and whose fuel supply has *not* run out? This would be the consequence if the switches failed first. At the end of vigorous exercise, when people say they can't do any more, does the muscle biopsy ever show that they still have reserves of phosphocreatine?

Hultman: After vigorous exercise in normal subjects you always find very low phosphocreatine values, when the exercise load is higher than that corresponding to the subject's $\dot{V}o_2$ max. Along with that you find a high lactate content in the muscle (Harris et al 1977). If the exercise load is slightly lower than the $\dot{V}o_2$ max you find very low glycogen values, if the subject is highly motivated to do the exhausting work. Lack of motivation of course changes the picture. We recently studied a patient who could not voluntarily continue an isometric contraction at 60% of MVC more than a few seconds, due to a subjective feeling of fatigue. Electrical stimulation of the muscles showed a completely normal response, with a contraction duration of normal length and ending with a phosphocreatine value close to zero and a high lactate content in the muscle.

Edwards: I showed earlier that the calculated total energy cost of an isometric contraction is less in a contraction at 90% MVC than in one at 50% MVC. In the same way, Dr Saltin and Dr Karlsson (1971) showed that glycogen depletion after exercise at 100% MVC is much less than after exercise at 50% MVC to fatigue.

Wilkie: That is not answering my question, which was: do you find people who can't contract their muscles any more while they still have fuel in them?

Edwards: I am saying that there *is* fuel left at the time that they have to stop because of fatigue.

Saltin: It is important to distinguish between power output and the type of fatigue you have been talking about, Professor Edwards, with isometric contractions. In situations where subjects run out of fuel, after exercising for 2 or 2½ hours all the fibres are depleted of glycogen. When we ask them to make a maximum voluntary contraction, it is not significantly reduced from control. This tells us that the switch mechanism is still functioning, and is not the limiting factor.

Dawson: In all the *in vitro* studies that I am aware of, I know of no case in which metabolic changes were being investigated and the conclusion of the study was that fatigue is *not* due to metabolic causes. That includes our own studies. Conversely, and still dealing with *in vitro* experiments, I have never read any papers where people have been looking for the cause of fatigue in the activation of contraction and have concluded that fatigue *wasn't* in the activation of contraction.

Edwards: It all depends whether you define fatigue in terms of force or power, and whether you are talking about the whole body or isolated fibres.

Dawson: I am talking about isolated fibres, and about force. Moreover, these different studies are done under remarkably similar conditions. Our studies are very similar to others on frog muscle in which the main conclusion is that fatigue results from changes in the activation of contraction (e.g. Eberstein & Sandow 1963, Grabowski et al 1972, Vergara et al 1977, Nassar-Gentina et al 1978). One can conclude from this that perhaps no one has done the right experiments, but also that the differences in opinion about the cause of fatigue result to a large degree simply from the different measurements we make, rather than from actual differences in the mechanism of fatigue under similar circumstances. For this reason, I have never myself been as interested in the actomyosin ATPase coming to a halt as a mechanism of fatigue, as in theories like the one that Dr Wiles and others have put forward—that the metabolic changes that occur when muscles contract have an effect on the activation of further contraction. This hypothesis accommodates most of the observations in the literature. In other words, we are all right!

Edwards: That brings us to Lars Hermansen's hypothesis that the accumulation of protons within the cell may result in impaired actomyosin activation. The question is of how we explore that hypothesis, to see whether this is a carefully regulated process which might have physiological significance or an unspecific result of the accumulation of protons.

Hultman: Dr Wiles compared two muscles stimulated during anoxia to fatigue, one a normal muscle and one from a patient with myophosphorylase deficiency. The action potential decreased to practically zero in both muscles during stimulation but returned in the normal muscle during the period of anoxic recovery. In the myophosphorylase-deficient patient the action potential in the muscle did not return during the recovery period. What is the reason for this difference? It cannot be the lactate or H^+ accumulation, as this will be higher in the normal muscle. The ATP content is high in both the muscles. The difference could be that in normal muscle, glycogenolysis and glycolysis can occur, making ATP turnover possible, while these processes are inhibited in the phosphorylase-deficient muscles.

As an action potential can be produced only in membranes with a normal resting potential, this normal potential is a prerequisite for the activation of the muscle. The resting potential can be kept normal only when the ATPase activity in the membrane is preserved. An inhibition of membrane ATPase by the accumulation of ADP and P_i due to a decreased rate of rephosphorylation of ADP in the phosphorylase-deficient subject could be the reason for the difference seen in the action potential. This could be interpreted to mean that fatigue is a decrease in membrane ATPase activity produced by product inhibition and inhibiting further stimulation of the contractile system of the muscle at the membrane level.

Wilkie: I would like to make a comment on what Joan Dawson said. My reason for being sceptical about some experiments by the 'activation' school is that caffeine is used as an experimental tool in many of them, and isometric recording is used in all of them. The experiments frequently don't make the important distinction between developing tension in the course of a normal contraction or as a result of going into an irreversible contracture. In the contracture, if you release the muscle even slightly its tension falls, it can't do any work and it is in a totally different state from normal contraction. In many cases the contracture is irreversible. One must therefore be very careful in interpreting such experiments as showing that the 'switches' fail.

Edwards: In studies by Donald Wood (1978) on normal muscles and dystrophic muscles that had been skinned there was little fatigue of the actomyosin contractile mechanism itself, providing that ATP and Ca^{2+} were present.

Wilkie: No one has examined the role of the products of ATP hydrolysis, although their concentrations are increased more than 10-fold. This is what is relevant to fatigue in intact muscle. Lars Hermansen has done experiments in which he decreased pH.

Hermansen: Yes. Single muscle fibres can be isolated in silicone oil and their sarcolemmas mechanically removed in an infused bubble of relaxing solution, as described by Donaldson & Kerrick (1975). In our studies (Donaldson et al 1978) we showed that the maximal Ca^{2+}-activated tension was lower at pH 6.5 than at pH 7.0. To me that indicates that the hydrogen ion concentration, in one way or another, must affect tension development. It means that in situations where there is a large fall in intracellular pH, tension develop is eventually affected. This doesn't mean that tension development couldn't fall for other reasons.

As a general point, I should like to stress that we have mostly discussed isometric exercise in this symposium. The studies I referred to in my paper involved dynamic exercise. There is a big difference between the two so far as the pattern of motor neuron activity is concerned. In isometric exercise motor neurons are firing continuously while in dynamic exercise they fire for a certain period of time, then they stop and then firing comes back again. So the two situations cannot really be compared.

REFERENCES

Brodal A 1973 Self-observation and neuro-anatomical considerations after a stroke. Brain 96:675-694

Donaldson S, Kerrick W 1975 Characterization of the effects of Mg^{2+} on Ca^{2+}- and Sr^{2+}-activated tension generation of skinned skeletal muscle fibres. J Gen Physiol 66:427-444

Donaldson SK, Hermansen L, Bolles L 1978 Differential, direct effects of H^+ on Ca^{2+}-activated force of skinned fibers from soleus, cardiac and adductor magnus muscles of rabbits. Pflügers Arch Eur J Physiol 376:55-65

Edwards RHT, Nordesjö L-O, Koh D, Harris RC, Hultman E 1971 Isometric exercise—factors influencing endurance and fatigue. Adv Exp Med Biol 11:357-360

Grabowski W, Lobsiger EA, Lüttgau HCh 1972 The effect of repetitive stimulation at low frequencies upon electrical and mechanical activity of single muscle fibres. Pflügers Arch Eur J Physiol 334:222-239

Eberstein A, Sandow A 1963 Fatigue mechanisms in muscle fibres. In: Gutmann E, Hník P (eds) The effect of use and disuse on neuromuscular functions. Elsevier, Amsterdam

Hallett M, Khoshbin S 1980 A physiological mechanism of bradykinesia. Brain 103:301-314

Harris RC, Edwards RHT, Hultman E, Nordesjö L-O, Nylind B, Sahlin K 1976 The time course of phosphorylcreatine resynthesis during recovery of the quadriceps muscle in man. Pflügers Arch Eur J Physiol 367:137-142

Harris RC, Sahlin K, Hultman E 1977 Phosphagen and lactate contents of m. quadriceps femoris of man after exercise. J Appl Physiol 43:852-857

McDonald WI, Sears TA 1970 The effects of experimental demyelination on conduction in the central nervous system. Brain 93:583-598

Merton PA 1970 The sense of effort. In: Breathing: Hering-Breuer centenary symposium. Churchill, London (Ciba Found Symp) p 207-211

Nassar-Gentina V, Passonneau JV, Vergara JL, Rapoport SI 1978 Metabolic correlates of fatigue and recovery from fatigue in single frog muscle fibres. J Gen Physiol 72:593-606

Pedley TJ (ed) 1977 Scale effects in animal locomotion. Academic Press, London

Saltin B, Karlsson J 1971 Muscle glycogen utilization during work of different intensities. Adv Exp Med Biol 11:289-299

Sherrington CS 1900 In: Schafer EA (ed) Textbook of physiology. Pentland, Edinburgh & London, vol 2:1002-1013

Vergara JL, Rapoport SI, Nassar-Gentina V 1977 Fatigue and posttetanic potentiation in single muscle fibres of the frog. Am J Physiol 232(3):C185-C190

Wilkie DR 1974 The efficiency of muscular contraction. J Mechanochem Cell Motility 2:257-267

Wood DS 1978 Human skeletal muscle: analysis of Ca^{2+} regulation in skinned fibers using caffeine. Exp Neurol 58:218-230

Chairman's summing-up

R. H. T. EDWARDS

Department of Human Metabolism, University College London School of Medicine, University Street, London WC1E 6JJ, UK

I now have the task of summing up the symposium and attempting to give an overall view of human muscle fatigue. It has been a privilege to bring together friends from many different disciplines. We have covered much ground. We started off from viewpoints that were rather far apart, but we have enjoyed a remarkably free exchange of ideas and we shall none of us think of fatigue again in quite the same way.

We have covered the essential links in the chain of command for muscular contraction (Fig. 1) and the means by which force may fail, and have

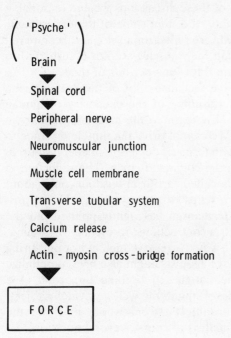

FIG. 1. Command chain in voluntary contraction of skeletal muscle. (From Edwards 1978.)

compared the function of human muscle *in vivo* with function in isolated muscle preparations. We have considered the identification of central fatigue and various examples of what it could represent in physiological terms. A useful concept has emerged from Brenda Bigland-Ritchie's work and her suggestion that central fatigue itself has to be subdivided into two types, one of which John Stephens has called relaxation, and which Lennart Grimby also considered as a form of central fatigue—that is, a lack of recruitment of motor units. Brenda Ritchie sought to separate from this a second type which was obligatory in that force could not be restored, by any means, however well motivated the subject, for even a brief period. This might represent an intrinsic feature of the central control system, possibly due to a reduction in afferent influences on motor unit firing frequencies. Central fatigue may thus be thought of as comprising both a motivational component and a reflex control component.

We next turned to peripheral fatigue. Here I should like to quote Pat Merton again, who in 1956 wrote 'Anyone who possesses a sphygmomanometer and an open mind can readily convince himself that the site of fatigue is in the muscles themselves'. Of course, fatigue can be demonstrated to be peripheral in certain circumstances. As we have repeatedly pointed out, the conclusion as to where the site of fatigue might be depends enormously on the type of muscular activity. At the end of these discussions we had reasonable agreement about that and the possibility, too, that some of the experiments that Pat Merton and we have been doing are rather unphysiological compared with pedalling bicycles. By that I mean that muscles are not accustomed to receiving high frequency trains of stimuli for long periods of time.

Another exciting aspect has been the fact that a lot of the Scandinavian work from which much of our understanding of the chemistry of human muscle comes has also shown how well integrated the mechanisms that we have been discussing are in practice. We cannot drive the muscle machine to any kind of limit which is recognizable in terms of depletion of energy sources if something fails more proximally in the command chain. The reduction in firing frequency that Pat Merton has called 'artificial wisdom', and about which Brenda Bigland-Ritchie spoke, helps to preserve excitation in prolonged isometric contraction. Lars Hermansen has rightly pointed out the difference between phasic and isometric contractions: in cycling or running, the phasic muscular activity might be a way to preserve high motor unit firing frequencies. The rotating involvement of agonists might also help to optimize muscle function, allowing 'metabolic' fatigue to become important. We assume that all these mechanisms are working in the well-motivated subject. The motivation of the Scandinavian subjects, from whom much of the metabolic information has been obtained, is of course beyond question!

We can learn from the study of patients with selected defects in the command chain (Fig. 1) or in energy metabolism. Patients with defects in the central nervous system allow the recognition of the relative importance of central and spinal or peripheral pathways in motor control. Patients with metabolic defects are important to our understanding of how muscle works in health. The studies that Mark Wiles has described help to bridge the gap in our understanding between alterations in energy supply and changes in membrane function as causes of fatigue.

If we now consider the muscle cell, Fig. 2 shows a three-dimensional representation of the relation between excitation/activation and energy supply as determinants of force. We know from David Jones's work that there are changes in contractile properties that take place with muscular activity. The alterations in twitch and relaxation characteristics are happening continuously during activity, even when there is no loss of force. I think it is worth distinguishing these from fatigue, even though they may influence the overall muscle performance. Fatigue is the point when performance fails, as with a chain breaking—that is, an all-or nothing phenomenon. There may be a failure of the supply of energy at a sufficient rate to meet demands or a failure in the completeness of the excitation–contraction coupling mechanism, which has to be intact to sustain the demand for energy exchange and the required performance.

If we consider these factors in order (Fig. 2), the relationship between force and energy is something that has received attention in the past. Murphy (1966) described the relationship between the ATP content and the force in muscle that had been poisoned. Spande & Shottelius (1970) described the relationship between phosphocreatine content and force. Dr Dawson and Professor Wilkie have derived a linear relationship between rate of ATP hydrolysis and force. On the other side, if we consider excitation, the frequency–force relationship that is demonstrable in human muscle studied *in vivo* follows the same curve that Dr Hermansen showed us for calcium concentration and force, and in both these circumstances the steep part of the curve is capable of being shifted in one direction or the other, indicating that the efficacy of coupling between excitation or activation and force generation is variable.

The part of Fig. 2 which is most difficult is the plane representing energy and activation. All I can do is to try to draw this curve and to say that so long as excitation is subject to factors other than energy supply, this cannot be a simple linear relationship. The other factor is of course the accumulation of potassium ions or depletion of sodium ions that can result in failure of excitation at the membrane at high stimulation frequencies.

What happens in different forms of exercise is worthy of both further thought and more precise experiments. There might be a different course within this three-dimensional plot representing the 'collapse' of the chain with

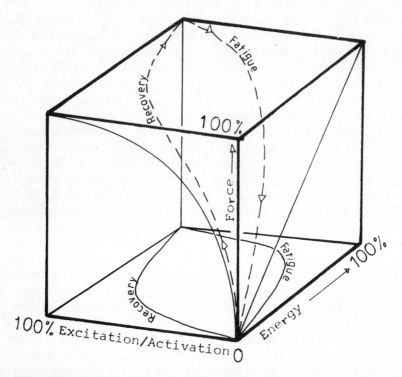

FIG. 2. Three-dimensional plot of possible relation between force and energy (ATP hydrolysis, Dawson et al 1978) and excitation/activation (frequency:force curve, Edwards et al 1977). This representation is intended to emphasize our present lack of quantitative information about the precise interrelationships between the variables. Recovery may follow a different path, as a result of the different time courses of recovery processes. Both fatigue and recovery pathways may vary according to the type of muscular activity.

different forms of exercise. It is also likely that the recovery processes follow very different time courses. As we have heard, there are clear differences in the time courses of the recovery of excitation, and of phosphocreatine and maximum force (fairly rapid), and in the removal of lactate (moderately slow), and these are clearly different from the slow recovery of low frequency fatigue. I believe that a study of recovery would be valuable in separating these elements, which may appear linked during the process leading to fatigue. Generally speaking, it is not necessarily so clever to show that a muscle doesn't work, which is in the end what fatigue is. I am more interested in understanding how it can recover and work again. I hope it will be possible in the next few years to have information on the relationship between energy and excitation/activation. What is clear is that the energy requirements of activation itself are quite large, from the experiments of Homsher et al (1972) in which frog muscle fibres were stimulated when there was no overlap of the actin and myosin, so that force was reduced to zero but the measured heat production was 30% of that measured in a twitch at resting length.

Another enigma is the energy exchange associated with excitation. As Mark Wiles has pointed out, a query exists over anaerobic recovery in patients with myophosphorylase deficiency, and the failure of the action potential to recover at a time when there appears to be sufficient resources for anaerobic recovery and there is adequate opportunity for the recovery of the action potential in normal subjects. As I understand it, the energy requirements of action potential generation and propagation are extremely small, so we are dealing with two processes, one that might have a large energy requirement and the other a very small one.

In conclusion, my view of fatigue is analogous to the way Barcroft (1934) described exercise in his *Features in the architecture of physiological function*—a book that has influenced me over the years. He said that exercise is an example of his general aphorism: every adaptation is an integration. I suppose that in a way we are talking about a disintegration, the collapse of muscle function when fatigue occurs. I hope that what we have gained from this symposium is not only an idea of the ways in which we may identify the site of fatigue in any particular muscle activity but also information that may help us to understand, and wonder at, the efficacy of the integration that enables muscle function to continue normally.

REFERENCES

Barcroft J 1934 Features in the architecture of physiological function. Cambridge University Press, London, p 187

Dawson MJ, Gadian DG, Wilkie DR 1978 Muscular fatigue investigated by phosphorus nuclear magnetic resonance. Nature (Lond) 274:861-866

Edwards RHT 1978 Physiological analysis of skeletal muscle weakness and fatigue. Clin Sci Mol Med 54:463-470

Edwards RHT, Young A, Hosking GP, Jones DA 1977 Human skeletal muscle function: description of tests and normal values. Clin Sci Mol Med 52:282-290

Homsher E, Mommaerts WFHM, Ricchiuti NV, Wallner A 1972 Activation heat, activation metabolism and tension-related heat in frog semitendinosus muscles. J Physiol (Lond) 220:601-625

Merton PA 1956 Problems of muscular fatigue. Br Med Bull 12:219

Murphy RA 1966 Correlations of ATP content with mechanical properties of metabolically inhibited muscle. Am J Physiol 211:1082-1088

Spande JI, Shottelius BA 1970 Chemical basis of fatigue in isolated mouse soleus muscles. Am J Physiol 219:1490-1495

Index of contributors

Subject index